Energiepolitik und Klimaschutz
Energy Policy and Climate Protection

Herausgegeben von
L. Mez, Berlin, Deutschland
A. Brunnengräber, Berlin, Deutschland

Weltweite Verteilungskämpfe um knappe Energieressourcen und der Klimawandel mit seinen Auswirkungen führen zu globalen, nationalen, regionalen und auch lokalen Herausforderungen, die Gegenstand dieser Publikationsreihe sind. Die Beiträge der Reihe sollen Chancen und Hemmnisse einer präventiv orientierten Energie- und Klimapolitik vor dem Hintergrund komplexer energiepolitischer und wirtschaftlicher Interessenlagen und Machtverhältnisse ausloten. Themenschwerpunkte sind die Analyse der europäischen und internationalen Liberalisierung der Energiesektoren und -branchen, die internationale Politik zum Schutz des Klimas, Anpassungsmaßnahmen an den Klimawandel in den Entwicklungs-, Schwellen- und Industrieländern, die Produktion von biogenen Treibstoffen zur Substitution fossiler Energieträger oder die Probleme der Atomenergie und deren nuklearen Hinterlassenschaften.

Die Reihe bietet empirisch angeleiteten, quantitativen und international vergleichenden Arbeiten, Untersuchungen von grenzüberschreitenden Transformations- und Mehrebenenprozessen oder von nationalen „best practice"-Beispielen ebenso ein Forum wie theoriegeleiteten, qualitativen Untersuchungen, die sich mit den grundlegenden Fragen des gesellschaftlichen Wandels in der Energiepolitik und beim Klimaschutz beschäftigen.

Herausgegeben von
PD Dr. Lutz Mez
Freie Universität Berlin

PD Dr. Achim Brunnengräber
Freie Universität Berlin

Jan Beermann

Urban Cooperation and Climate Governance

How German and Indian Cities
Join Forces to Tackle Climate Change

Jan Beermann
Freie Universität Berlin
Forschungszentrum für Umweltpolitik
Deutschland

Dissertation Freie Universität Berlin, 2016

D188

Gefördert durch das Stipendienprogramm der Deutschen Bundesstiftung Umwelt (DBU)

Energiepolitik und Klimaschutz. Energy Policy and Climate Protection
ISBN 978-3-658-17145-2   ISBN 978-3-658-17146-9   (eBook)
DOI 10.1007/978-3-658-17146-9

Library of Congress Control Number: 2017931969

Springer VS
© Springer Fachmedien Wiesbaden GmbH 2017
This work is subject to copyright. All rights are reserved by the Publisher, whether the whole or part of the material is concerned, specifically the rights of translation, reprinting, reuse of illustrations, recitation, broadcasting, reproduction on microfilms or in any other physical way, and transmission or information storage and retrieval, electronic adaptation, computer software, or by similar or dissimilar methodology now known or hereafter developed.
The use of general descriptive names, registered names, trademarks, service marks, etc. in this publication does not imply, even in the absence of a specific statement, that such names are exempt from the relevant protective laws and regulations and therefore free for general use.
The publisher, the authors and the editors are safe to assume that the advice and information in this book are believed to be true and accurate at the date of publication. Neither the publisher nor the authors or the editors give a warranty, express or implied, with respect to the material contained herein or for any errors or omissions that may have been made. The publisher remains neutral with regard to jurisdictional claims in published maps and institutional affiliations.

Printed on acid-free paper

This Springer VS imprint is published by Springer Nature
The registered company is Springer Fachmedien Wiesbaden GmbH
The registered company address is: Abraham-Lincoln-Str. 46, 65189 Wiesbaden, Germany

Acknowledgements

Special thanks go to everyone who supported me during my visits to cities in India and Germany. I am particularly grateful for the generosity you showed me both in terms of time and hospitality.

In particular I would like to thank my interviewees for providing me with the information that forms the basis of this study.

Moreover I am extremely grateful to my supervisors, Miranda Schreurs and Avi Gottlieb, for challenging me and providing valuable feedback and support, and to my colleagues and peers at the Freie Universität Berlin for their highly constructive and motivating feedback. Special thanks go also to Philipp Schönberger for testing the index system as a second rater.

I would also like to express my sincerest thanks to the Deutsche Bundesstiftung Umwelt (DBU) for offering me generous and flexible funding, and for the training in how to present my research to an interdisciplinary audience. In particular I would like to thank Hedda Schlegel-Starmann for her great support and supervision.

Many thanks also to the Deutsche Forschungsgemeinschaft (DFG), the Indian-European Multi-Level Climate Governance Research Network and its coordinator Kirsten Jörgensen and for providing the funding for my research stay in India and for giving me the opportunity to present and discuss my research on several occasions.

I would also like to thank my friends and family for keeping me motivated, for listening to my concerns and for showing interest in the progress of my work.

My greatest thanks go to Erica, my salvation in hard times, whose support for this study extended far beyond countless hours of proofreading and editing.

I dedicate this work to Erica and our children, Tilda and Erik Anton.

Table of Contents

1 Introduction 17
   1.1 Background 17
      1.1.1 The Growing Focus on Cities as a Proactive Force in Global Climate Governance 18
      1.1.2 Emerging Transnational Urban Co-operation towards a Low-Carbon Transition 21
   1.2 Purpose of the Study 23
   1.3 Research Questions and Hypotheses 25
   1.4 Organisation of the Study 26

2 State of Research and Key Debates 29
   2.1 Urban Climate Governance 29
   2.2 Urban Climate Governance in Germany and India 32
   2.3 Urban North-South Cooperation 36
      2.3.1 State Actor-driven Transnational Urban Cooperation 36
      2.3.2 Private Actor-driven Transnational Urban Cooperation 38
      2.3.3 Urban Cooperation in Transnational Municipal Networks 39
   2.4 Transnational Urban Cooperation in Germany and India 40
      2.4.1 Transnational Urban Cooperation in Germany 41
      2.4.2 Transnational Urban Cooperation in India 43
      2.4.3 Urban Cooperation between German and Indian Cities 45

3 Theoretical Framework 47
   3.1 Policy Transfer Theory 47
   3.2 Transnational Climate Governance Networks 50

3.3	New Institutionalism	52
3.4	Policy Entrepreneur Concept	55
3.5	Social Capital Theory	57

4 Methodology ... 61

- 4.1 Addressing the Northern Bias in Urban Research ... 61
  - 4.1.1 Assumed Incommensurability ... 61
  - 4.1.2 Northern Bias Reflected in Urban Climate Change Research ... 62
  - 4.1.3 Focus on Urban Processes and City Connections as a Way Forward ... 62
- 4.2 Comparative Case Study Analysis ... 63
  - 4.2.1 Qualitative Case Study Analysis ... 63
  - 4.2.2 Multiple Case Study Analysis ... 64
- 4.3 Research Approach ... 65
- 4.4 Development of Index System to Operationalise Dependent and Independent Variables ... 66
  - 4.4.1 Approach ... 66
  - 4.4.2 Index System ... 67
- 4.5 Case Selection ... 72
- 4.6 Data Collection ... 73

5 Within Case Analysis ... 75

- 5.1 Case Study Pune – Bremen I: Developing Decentralised Wastewater Treatment Systems ... 75
  - 5.1.1 Pune-Bremen History: A Longstanding Bottom-up Partnership in Sustainable Development ... 75
  - 5.1.2 Partnership Project DEWATS: Content and Process ... 81
  - 5.1.3 Project DEWATS: Outcomes ... 84
  - 5.1.4 Partnership Project DEWATS: Explanatory Factors ... 90
- 5.2 Case Study Pune – Bremen II: Transferring the Bremen Tramway System to Pune ... 113
  - 5.2.1 Partnership Project Tramway: Content and Process ... 113

5.2.2	Partnership Project Tramway: Outcomes	117
5.2.3	Partnership Project Tramway: Explanatory Factors	121
5.3	Case Study Nashik – Hamburg: Reducing Emissions through Waste-to-Energy	139
5.3.1	Partnership Project W2E: Content and Process	139
5.3.2	Partnership Project W2E: Outcomes	143
5.3.3	Partnership Project W2E: Explanatory Factors	149
5.4	Case Study Nagur – Freiburg: The Local Renewables Model Community Network	167
5.4.1	Partnership Project LRMCN: Content and Process	167
5.4.2	Partnership Project LRMCN: Outcomes	172
5.4.3	Partnership Project LRMCN: Explanatory Factors	179

6 Cross-Case Comparison: Testing the Research Hypotheses ... 199

6.1 Results from Comparative Case Study Analysis ... 199

6.2 Comparing Partnership Outcome: Mixed Success and Differing Strengths and Weaknesses ... 200

6.3 Explanatory Factors in a Comparative Perspective ... 205

    6.3.1 Knowledge Exchange Strategy ... 205

    6.3.2 Linkages between the Partnership Project and State Institutions ... 209

    6.3.3 Partnership Entrepreneurs ... 213

    6.3.4 Partnership Social Capital ... 215

6.4 Synopsis of Comparative Analysis ... 219

7 Discussion and Conclusions ... 223

7.1 Key Factors Determining the Success and Scope of Urban North-South Cooperation ... 223

    7.1.1 The Potential of Multi-Level Governance Coordination to Address the Gap between Bottom-up and Top-Down Approaches ... 223

7.1.2 Pursuing Intrinsic Interests and Co-benefits to Strengthen
Mutuality in Urban North-South Partnerships............................ 233
7.2 Academic Implications 236

8 References 245

9 Appendix 255
9.1 List of Interviews 255
9.2 Summary 259
9.3 Zusammenfassung 261

Figure and Tables

Figure 1:	Tramway alignment plan for Pune and Pimpri Chinchwad; source: Sakhalkar (2007)	116
Table 1:	Typology of Transnational Climate Change Governance Networks; source: Andonova, Betsill and Bulkeley, 2009	51
Table 2:	Partnership Project DEWATS: Findings from success index	84
Table 3:	Partnership Project DEWATS: Findings from Index Knowledge Exchange Strategy	91
Table 4:	Partnership Project DEWATS: Findings from Index State Institutionalisation	96
Table 5:	Partnership Project DEWATS: Findings from Index Partnership Entrepreneur	100
Table 6:	Partnership Project DEWATS: Findings from Index Partnership Social Capital	106
Table 7:	Partnership Project Tramway: Findings from success index Implementation Achievements	117
Table 8:	Partnership Project Tramway: Findings from Index Knowledge Exchange Strategy	122
Table 9:	Partnership Project Tramway: Findings from Index State Institutionalisation	128
Table 10:	Partnership Project Tramway: Findings from Index Partnership Entrepreneur	132

Table 11:	Partnership Project Tramway: Findings from Index Partnership Social Capital	135
Table 12:	Partnership Project Waste to Energy: Findings from success index	144
Table 13:	Partnership Project Waste-to-Energy: Findings from Index Knowledge Exchange Strategy	150
Table 14:	Partnership Project Waste-to-Energy: Findings from Index State Institutionalisation	158
Table 15:	Partnership Project Waste-to-Energy: Findings from Index Partnership Entrepreneur	162
Table 16:	Partnership Project Waste-to-Energy: Findings from Index Partnership Social Capital	164
Table 17:	Partnership Project LRMCN: Findings from success index	172
Table 18:	Partnership Project LRMCN: Findings from Index Knowledge Exchange Strategy	180
Table 19:	Partnership Project LRMCN: Findings from Index State Institutionalisation	186
Table 20:	Partnership Project LRMCN: Findings from Index Partnership Entrepreneur	190
Table 21:	Partnership Project LRMCN: Findings from Index Partnership Social Capital	193
Table 22:	Results of the Index System Scoring (Success Index highlighted)	200
Table 23:	Success Index: Results of the Indicator Scoring	201
Table 24:	Indicator Capacity Building: Scoring of the Indian and German Cities	202

Table 25:	Results of the Index System Scoring (Index 1 "Exchange Strategy" highlighted)	206
Table 26:	Index 1 Exchange Strategy: Results of the Indicator Scoring	207
Table 27:	Indicator Intrinsic Interests: Scoring of the Indian and German Cities	208
Table 28:	Results of the Index System Scoring (Index 2 "State Institutionalisation" highlighted)	209
Table 29:	Index 2 "State Institutionalisation": Results of the Indicator Scoring	211
Table 30:	Results of the Index System Scoring (Index 3 "Partnership Entrepreneur" highlighted)	213
Table 31:	Indicators Access to Policy Networks: Scoring of the Indian and German Cities	215
Table 32:	Results of the Index System Scoring (Index 4 "Social Capital" highlighted)	216
Table 33:	Index 3 "Partnership Social Capital": Results of the Indicator Scoring	217
Table 34:	Results of the Index System Scoring (Selected Indexes and Indicators)	220
Table 35:	Results of the Index System Scoring (Selected Indicators)	221
Table 36:	List of Interviews	258

Abbreviations

AFG	Association of the Friends of Germany
ASEM	Advisory Service for Environmental Management
AWO	Arbeiterwohlfahrt
BORDA	Bremen Overseas Research and Development Association
BMU	Bundesministerium für Umwelt, Naturschutz und Reaktorsicherheit – German Federal Ministry for the Environment, Nature Conservation and Nuclear Safety
BMZ	Bundesministerium für wirtschaftliche Zusammenarbeit und Entwicklung – German Federal Ministry for Economic Cooperation and Development
BRT	Bus Rapid Transit
BSAG	Bremen Straßenbahn AG
C2C	city-to-city cooperation
C40	C40 Cities Climate Leadership Group
CCP	Cities for Climate Protection
CDD	Consortium for DEWATS Dissemination
CDM	Clean Development Mechanism
CIM	Centre for International Migration and Development
CMP	Comprehensive Mobility Plan
CO_2	carbon dioxide
COP	Conference of Parties
CTB	Consult Team Bremen
DC	decentralized cooperation
DEWATS	decentralized wastewater treatment systems
DPR	Detailed Project Report
GPC	Global Protocol for Community-Scale Greenhouse Gas Emission Inventories
GDP	Gross Domestic Product
GHG	greenhouse gas
GIZ	Deutsche Gesellschaft für Internationale Zusammenarbeit GmbH
GTZ	Deutsche Gesellschaft für Technische Zusammenarbeit
HWC	Hamburg Water Cycle®

ICI	International Climate Initiative
ICLEI	ICLEI – Local Governments for Sustainability
IIBR	Institute of International Business and Research
IGEP	Indo-German Environment Partnership
IO A21	International Office Agenda 21
JICA	Japanese International Cooperation Agency
JnNURM	Jawarhalal Nehru National Urban Reform Mission
LAFEZ	Landesamt für Entwicklungszusammenarbeit
LRMCN	Local Renewables Model Communities Network
Mahagenco	Maharashtra Generation Company
MAM	Maharashtra Arogya Mandal
MLG	Multi-Level Governance
MNRE	Ministry of New and Renewable Energies
MoU	Memorandum of Understanding
MSW	municipal solid waste
Nagpur MC	Nagpur Municipal Corporation
NAPCC	National Action Plan on Climate Change
Nashik MC	Nashik Municipal Corporation
NGO	non-governmental organisation
NKI	National Climate Protection Initiative
OD	origin-destination
OECD	Organisation for Economic Development and Cooperation
PCMC	Pimpri Chinchwad Municipal Corporation
PE	partnership entrepreneur
PEARL	Peer Experience and Reflected Learning
RGRE	Rat der Gemeinden und Regionen Europas - Council of European Municipalities and Regions
PMC	Pune Municipal Corporation
SC	social capital
TMN	transnational municipal network
UCLG	United Cities and Local Governments
UK	United Kingdom
UN	United Nations
UNDARP	United Socio Economic Development and Research Programme
UNFCCC	United Nations Framework Convention on Climate Change
USA	United States of America
U.S.	United States
W2E	waste-to-energy
WRE	World Resources Institute

1 Introduction

1.1 Background

Global climate change is one of the most severe challenges that humankind faces in the 21st century. Global levels of anthropogenic greenhouse gas (GHG) emissions are increasing and the resulting warming of the planet threatens the livelihood of current and future generations. The developing countries and emerging economies of the Global South in particular are already feeling the consequences of climate change, such as rising sea levels, tropical storms, increased flooding and extended droughts. The industrialised countries of the Global North have also been more regularly exposed to extreme weather events (IPCC, 2014a). Most countries have now recognised man-made climate change as a serious global threat and climate protection has been widely acknowledged as a crucial political, socio-economic and technological task for the coming decades.

Since the early 1990s the international community has started to address this problem under the United Nations Framework Convention on Climate Change (UNFCCC). At annual Conferences of Parties (COPs) to the UNFCCC the delegates conduct negotiations on how to share the burden of reducing global GHG emissions and the fair distribution of emission rights. The results of these formal international negotiations between nation states have been criticised as unsatisfactory and not sufficient to keep global warming within a limit of two degrees Celsius that is proposed by scientists as necessary to prevent even higher risks of catastrophic climate change (Biermann et al., 2012; Bulkeley and Newell, 2010; Prins et al., 2010). The Kyoto Protocol's binding emissions targets for many industrialised countries have been only weakly enforced. Since the completion of the first commitment period (2008-2012), delegates have been struggling to develop a comprehensive international agreement to replace the Kyoto Protocol, including the setting of more ambitious targets and involving more countries. At the COP 18 in Doha in November and December 2012 delegates merely agreed to extend the Kyoto Protocol until 2020, with even fewer countries committing to binding emission reduction targets than was the case during the first commitment period. Canada quit the treaty and Japan, Russia and New Zealand did not adopt any new targets for the second phase. The adoption of the Paris Agreement on December 12th 2015 at the COP 21 in Paris has therefore been hailed as a landmark in international climate negotiations. For the first time, nearly all the world's countries agreed to commit to global climate protection efforts, setting the target to keep global warming "well below 2 °C above pre-industrial levels and pursuing efforts to limit the temperature increase to 1.5 °C above pre-industrial levels" (United Nations Framework Convention on Climate Change, 2015, 2). The agreement will become binding, if 55 countries making up 55% of global GHG emissions ratify it. By 2020, all countries that ratify the agreement will set up national climate protection strategies, which will be reviewed every five years, starting in 2023. One of the Paris Agreement's major achievements is its inclusion of the United States (US) which refused to ratify the Kyoto Protocol, and large developing and emerging economies, such as China, India and Brazil.

A key barrier that has for a long time prevented a formal international agreement has been the divide in the negotiations between industrialised countries from the Global North and developing countries from the Global South.[1] Contested issues have included whether developing countries should also commit to binding emissions targets and how to financially compensate highly vulnerable developing countries and island states for the negative impacts of climate change, for which they have little responsibility due to their relatively low historical and per capita emissions levels. Bulkeley and Newell (2010, 31) point out that the conflict between the Global North and the Global South around equity and justice is one of the key challenges in attempts to address global warming: "The broader historical and contemporary features of the unequal relationship between the developed and developing world run through virtually all aspects of climate governance."

1.1.1 The Growing Focus on Cities as a Proactive Force in Global Climate Governance

Because of the slow progress in the negotiations between nation states there has been increasing attention on the benefits of voluntary climate action at the subnational level. Cities are, in particular, regarded by many decision makers and researchers as important alternative arenas for climate mitigation and adaptation efforts. Bulkeley and Betsill (2013, 136) conclude that cities have become the

---

1 Many works from the field of post-colonial studies criticise the use of normative categories such as "rich" or "developed" cities from the "Global North" and "poor" or "underdeveloped" cities from the "Global South" (e.g. Conell 2007; McFarlane 2010). McFarlane (2010, 726) however acknowledges the political merits of using the term "Global South" and emphasises that "categories such as these cannot be simply written away." Supporting this argumentation, in particular with regard to the common but differentiated responsibilities of how countries address global climate change, the categories of "Global South" and "Global North" are also used in this study.

bearers of hope in international climate governance: "Far from being a little-known concern amongst a minority of municipalities, the city now looms large on the international climate change agenda." Whereas cities are not yet formally involved in the official UNFCCC negotiations, city representatives and transnational municipal networks have actively participated in many side events at the COPs. On November 21st 2013 the first ever "Cities Day" was held within the climate negotiations at the COP 19 in Warsaw. United Nations (UN) Secretary General Ban Ki-moon highlighted that "Cities are central in tackling climate change. They are proving grounds for our efforts in ensuring a low carbon future that benefits people and the planet" (ICLEI, 2013). The Paris Agreement recognises and highlights the role of cities and other non-state actors, welcoming "the efforts of all non-Party stakeholders to address and respond to climate change, including those of civil society, the private sector, financial institutions, cities and other subnational authorities." (United Nations Framework Convention on Climate Change, 2015, 2)

In fact, there are several reasons to consider cities as crucial sites for global climate protection. Already today more than half of the world's population lives in cities and according to United Nations urbanisation projections by 2050 66% of mankind will live in urban areas (United Nations, 2014). Calculations about cities' contributions to global GHG emissions vary widely, depending on amongst other factors whether both direct and indirect emissions are taken into account. There is also an unresolved debate about whether cities generate higher per capita GHG emissions compared to rural areas and should therefore be blamed as major contributors to climate change (Dhakal, 2009; Dodman, 2009; Satterthwaite, 2008, 2009).

The urban-rural divide is becoming increasingly blurred with cities and their surrounding rural areas often being considered as metropolitan entities. It may therefore be more insightful to regard the joint impact of local and regional governments which according to a study by the United Nations Development Programme (2009) are able to influence 50-80% of global GHG emissions. Cities have core functions in the multi-level governance of climate change. They are crucial sites for the implementation of national climate change strategies and for experimenting with tailored local responses to climate change mitigation and adaptation. As hubs for technological and social transformation, urban centres have huge potential to shape low carbon development pathways (Kamal-Chaoui and Robert, 2009). In fact, both in the Global North as well as increasingly in the Global South, cities have become major arenas for climate policy innovation and low carbon development. Cities promote energy-saving strategies, set their own renewable energy targets, improve sustainable transport infrastructure and enhance climate-friendly city planning. Urban areas often even serve as "first responders" to climate change in their countries (Rosenzweig, Solecki, Hammer and Mehrotra, 2010, 909) and many local governments commit to GHG emissions reduction targets that meet or even exceed the commitments of national governments (Martinot, 2011; Schreurs, 2008). This has been especially the case in the United States and Japan where communities, cities and provinces/states have moved ahead by introducing their own climate and renewable energy targets in the absence of ambitious policies at the federal level (Krause, 2011; Schreurs, 2010).

Also in Germany, a global frontrunner in renewable energy deployment and climate mitigation approaches, many cities have engaged in climate protection for more than 25 years and several local governments have adopted local emission reduction targets that match or exceed the German government's commitments (Deutsches Institut für Urbanistik, 2011). Similarly, in other European countries, such as for example Sweden (e.g. Stockholm, Malmo, Gothenburg and Växjö), Denmark (e.g. Copenhagen and Samsø), Great Britain (e.g. London, Brighton and Edinburgh), France (e.g. Dunkirk and Besancon), Switzerland (e.g. Basel, Zurich and Geneva) and Spain (e.g. Barcelona and El Hierro) a large number of cities, municipalities and islands have been strongly committed to fostering local climate protection and sustainable development. Many cities are active in promoting local renewable energy development. This is exemplified by the prominent 100% Renewable Energy Communities and Regions movement, in which local governments commit to pursuing a local supply fuelled entirely by renewable energy sources. This movement has been particularly strong in Germany, other European countries and Japan and is now gaining popularity in cities around the world (Beermann, 2009; Beermann and Tews 2015; Martinot, 2011).

Compared to cities from the Global North, the development of urban climate strategies in the Global South is a more recent phenomenon. In emerging economies, such as India, China, and South Africa and in Latin America the number of cities engaged in local climate protection has steadily risen over the last decade (Aylett, 2010; Castán Broto and Bulkeley, 2013; Dhakal, 2009; Martinot, 2011; Revi, 2008; Qi, Ma, Zhang and Li, 2008; Sharma and Tomar, 2010; Yuen and Kong, 2009). Prominent examples include Curitiba (Brazil), a model city for Bus Rapid Transit systems (Campbell, 2012), Bangkok's (Thailand) light rail and underground rail systems (Yuen and Kong, 2009), Rizhao's (China) solar energy deployment (Schreurs, 2010) and Delhi's (India) efforts in reducing its carbon footprint (Sharma and Tomar, 2010).

1.1.2 Emerging Transnational Urban Co-operation towards a Low-Carbon Transition

Alongside aligning their climate and low carbon strategies closely to their specific contexts and needs, many cities are also collaborating and exchanging knowledge and experiences with other cities.

The benefits of urban international cooperation between cities on sustainable development have been internationally recognised since the UN Conference on Environment and Development in Rio de Janeiro in 1992 (the so-called "Earth Summit"). The conference's action programme, Agenda 21, highlights local governments as key actors in the implementation and diffusion of the concept of sustainable development worldwide:

> Because so many of the problems and solutions being addressed by Agenda 21 have their roots in local activities, the participation and cooperation of local authorities will be a determining factor in fulfilling its objectives. Local authorities construct, operate and maintain economic, social and environmental infrastructure, oversee planning processes, establish local environmental policies and regulations, and assist in implementing national and subnational environmental policies. As the level of governance closest to the people, they play a vital role in educating, mobilizing and responding to the public to promote sustainable development. (United Nations Conference on Environment and Development, 1992)

The Agenda 21 specifically calls for "increased levels of cooperation and coordination with the goal of enhancing the exchange of information and experience among local authorities" (ibid.).

Statz and Wohlfahrt (2010) explain that the Rio "Earth Summit" was a cornerstone that gave strong impetus for further international city cooperation on sustainable development. They point out that environmental protection and climate change in particular have become major issues in cities' international relations, played out in city twinning and decentralised cooperation, as well as in new forums such as transnational municipal networks.

The exact scope of transnational urban cooperation[2] can only be estimated. According to a popular figure published by the global city network United Cities and Local Governments (2007) (and cited amongst others by Bontenbal, 2009; Devers-Kanoglu, 2009 and van Ewijk and Baud, 2009), 70% of the world's cities are engaged in urban partnership activities. Campbell (2012, 9) estimates the scope of city-to-city exchange encompasses "thousands, perhaps tens of thousands, of visits each year across the globe". Many national governments also involve cities

---

2 In this study, 'transnational urban cooperation' describes any form of partnership involving state and/or non-state actors from two or more cities from different countries.

as partners in international development cooperation. Since the late 1980s most European countries as well as Canada and Japan have established national institutions to guide and support cities in their development engagement (Hafteck, 2003). In 2008 the European Commission introduced its first thematic funding program for decentralised cooperation[3], the "Non-State Actors and Local Authorities (NSA-LA)" (PLATFORMA, 2011).

In addition to these more established forms of urban cooperation a multiplicity of transnational municipal networks (TMNs) has emerged to foster local sustainable development and climate protection. They range from regional networks such as the Climate Alliance (largely German-speaking cities), Energy Cities (largely French-speaking cities), the Covenant of Mayors (largely European cities) and CITYNET (largely Asian cities), and thematically focused networks such as the Clean Air Initiative for Cities Around the World and the CIVITAS initiative for sustainable transport, to global networks covering millions of city inhabitants worldwide such as ICLEI – Local Governments for Sustainability (ICLEI), the C40 Cities Climate Leadership Group (C40) and the World Mayors Council on Climate Change. An indicator for the growing scope of TMNs is their growth in membership. One of the oldest and most prominent large-scale TMNs is ICLEI. Founded in 1990, ICLEI coordinates urban collaboration and knowledge exchange and advocates for cities' interests at the national and international level from its regional offices around the world. Today, ICLEI has become a global network of more than 1,000 cities representing more than 20% of the global urban population, with regional offices on all continents.[4] A second prominent TMN is C40, a network encompassing 83 large and mega cities from all over the world, which together make up about 8 percent of the world's population and 25 percent of global GDP. C40 was founded in 2005 and at least 45 of its member cities have published a local climate action plan.[5] A third TMN which has seen a very rapid increase in membership is the European Covenant of Mayors. Established in 2008, the Covenant of Mayors already includes more than 6,700 local and regional authorities that have committed to setting up local clean energy strategies and report on their

---

[3] 'Decentralised cooperation' is defined as sub-national North-South cooperation on sustainable development, usually between local governments and featuring exchange and support as its main activities (Hafteck, 2003, 333). While the term 'decentralised cooperation' is still used in practice, in more recent literature it has become more common to use the terms 'city-to-city cooperation' (Bontenbal & van Lindert, 2009) and 'municipal international cooperation' (van Ewijk & Baud, 2009). In comparison to 'decentralised cooperation' the two latter terms are broader and also include non-state actors' activities, however, both terms remain state-biased and define local governments as the main actors in urban cooperation (see also section 2.3.1). This study therefore proposes the more neutral term of 'transnational urban cooperation' as it does not privilege either state or non-state actors' contributions to cross-border partnerships between cities.

[4] http://www.iclei.org/ (19-02-2016)

[5] http://www.c40.org/cities (19-02-2016)

progress annually. The network covers more than 210 million citizens, mostly from European cities[6], and has been rated as one of the policies/actions likely to have the biggest impact on climate protection by 2020 in a research study commissioned by The Economist (2014).

The increasing interest in urban climate networking is also indicated by the fact that many established TMNs that generally have a wider thematic focus have made local climate action a priority. Examples are METREX – The Network of European Metropolitan Regions and Areas[7], Metropolis[8] and United Cities and Local Governments[9].

1.2 Purpose of the Study

This study offers a highly significant and topical contribution to the research field of urban climate governance by addressing several persisting research gaps.

As more and more cities experiment with new approaches to climate mitigation and low carbon development, experience of and knowledge about local GHG emissions reduction strategies is growing. Despite this increase in urban climate action and cooperation, there is still surprisingly little empirical and theoretical research available about city-level climate cooperation.

Campbell (2012, 10) states that learning among cities is generally a "blind spot" in urban development approaches as well as in academic literature. He finds that while cities are recognised as centres for innovation, "they have not been plumbed for their knowledge-exchange properties" (ibid., 9). Other researchers agree that local-level policy transfer remains a "black box" (Medearis and Dolowitz, 2013, 10; Wolman and Page, 2002, 478). Thus, there is a general demand to improve and conceptionalise the understanding of the drivers, institutional forms, processes and conditions for success and failure in urban cooperation.

With regard to the slow progress at the UNFCCC climate conferences the question of whether and under which conditions urban collaboration could serve as an alternative arena fostering direct learning and the diffusion of best practices, is highly significant. Is it possible to globally transfer urban low carbon knowhow, experiences and policies across city and national borders? Can experienced cities guide less advanced cities in their transition towards more sustainable and carbon-friendly development and what are the benefits for the "frontrunners" to

---

6 http://www.covenantofmayors.eu/index_en.html (19-02-2016)
7 http://www.eurometrex.org/ENT1/EN/ (19-02-2016)
8 http://www.metropolis.org/ (19-02-2016)
9 http://www.uclg.org/ (19-02-2016)

share their knowledge? What are the factors that trigger climate collaboration between cities and which are the crucial contextual conditions for the success and failure of urban cooperation on the mitigating of climate change?

Although the research community has recognised the importance of the topic of urban climate collaboration (Alber and Kern, 2008; Betsill and Bulkeley 2004, 2006; Bulkeley, 2006; Bulkeley and Betsill, 2013; Corfee-Morlot et al., 2009; Kern and Bulkeley, 2009), prominent research gaps remain. Three areas in particular require more attention. Firstly, there is a general lack of research on cooperation between cities from both the Global North and Global South. Secondly, there is a need for comparative research on different institutional designs of urban climate collaboration. In particular, urban partnerships set up and led by private actors have not been sufficiently explored. Thirdly, more research is required on the conditions and processes leading to success or failure in city cooperation (see chapter 2 for more detailed discussion of the existing literature).

The purpose of this study is to help address these knowledge and research gaps in urban South-North climate cooperation through an analysis of four Indian-German urban partnership projects. Via a comparative case study analysis the study investigates how German and Indian cities cooperate and learn from each other in the development of climate mitigation activities. The study sheds light on the specific conditions required for the successful exchange of knowledge, policies and technologies on urban climate action in spite of significant political and socioeconomic differences and geographical distance. The study strengthens current empirical insights and builds upon existing theories linked to the conditions, potential and limitations related to urban South-North climate cooperation. It introduces and compares distinct approaches to the designing of urban climate partnerships and discusses the respective benefits and shortcomings of bottom-up versus top-down development of transnational urban cooperation. As far as the author is aware, there currently exists no comparative analysis of German-Indian climate collaboration at the city level or any in-depth analysis comparing different institutional set-ups of urban North-South cooperation on low carbon development.

The findings have theoretical, methodological and practical applications. The study tests four research hypotheses on the conditions for success and failure of transnational urban partnerships derived from theoretical literature. As explained in-depth in chapter 3, the study draws on a set of theoretical concepts that jointly exert great explanatory power as they provide for distinct but complementary perspectives with regard to the analysis of transnational urban climate cooperation.

To operationalise the hypotheses, the methodological tool of an index system has been developed to assess the outcome of urban climate collaboration and specifically the conditions which lead to either the success or failure of such an endeavour. The index system is used to analyse the four German-Indian partnership projects selected. Based on the results of the projects' index scores the research hypotheses are tested, thereby evaluating and refining the theoretical concepts which the study draws upon.

The study also provides practical policy recommendations on how to improve and extend urban climate cooperation between cities from the Global South and North. The results of this study may be utilised to design and implement urban climate collaboration in a more effective manner and leverage the still largely untapped potential of transnational urban knowledge and policy transfer in the global transition towards low carbon development. The study aims to provide guidance on how to prepare, implement and evaluate urban cooperation and exchange of know-how, experiences and policies in addressing climate change in cities. Moreover, it offers recommendations on how to institutionalise urban partnerships within the multi-level governance of climate change and on how to facilitate local stakeholder participation, focusing specifically on the role of partnership entrepreneurs and the development of social capital within partnerships. The study also elaborates on how to pave the way for more equality and mutuality in international climate cooperation.

1.3 Research Questions and Hypotheses

The core of the study is an in-depth comparative case study analysis of four Indian-German urban partnership projects. The first two cases, the development of decentralised wastewater treatment systems and the attempt to transfer an urban tramway system from Germany to India are part of the long-term city cooperation between Pune and Bremen. The third case is a waste-to-energy project involving actors from Nashik and Hamburg, facilitated by the Deutsche Gesellschaft für Internationale Zusammenarbeit GmbH (GIZ). The fourth case is a partnership on renewable energy and energy efficiency development between Nagpur and Freiburg which was set up by the municipal network ICLEI (for more details on the methodology, including the case selection please see chapter 4).

The study addresses the following research questions via a comparative analysis of the four Indian-German urban partnerships:

What are the drivers, processes and outcomes of Indian-German urban partnerships on climate mitigation and low carbon development?

What are the specific conditions that lead to the success or failure of the setting up and implementation of Indian-German urban climate partnerships?

What are the benefits and limitations of bottom-up versus top-down approaches towards the designing of urban climate partnership initiatives?

To guide the analysis and position the study within the academic discourse, four research hypotheses are derived from existing theoretical literature (see chapter 3).

Hypothesis 1: A transnational urban partnership project is more likely to succeed if it follows a well-prepared knowledge exchange strategy.

Hypothesis 2: The more a transnational urban partnership project is institutionalised into the state system, the more likely the project is to succeed.

Hypothesis 3: A transnational urban partnership project is more likely to succeed if it is driven by engaged, persuasive and well-networked partnership entrepreneurs.

Hypothesis 4: The more social capital protagonists develop as part of the transnational urban partnership, the more likely the partnership project is to succeed.

The hypotheses are tested in the comparative case study analysis and, if necessary, refined or revised according to the empirical findings.

1.4 Organisation of the Study

Chapter 2 provides an overview of the current state of research and important debates in the academic literature on urban climate governance and transnational urban cooperation, with a specific focus on the German and Indian contexts. Chapter 3 introduces the study's theoretical framework, explaining how four research hypotheses are derived from the complementary theoretical concepts of policy transfer, transnational climate governance networks, New Institutionalism, policy entrepreneurship and social capital. In chapter 4 the methodology is outlined, detailing how to address and reduce Northern biases in urban research, rationales for selecting a comparative case study design and a "grounded" index system to operationalise the research hypotheses. Chapter 4 also explains the case selection and the methods of data collection (expert interviews and document analysis). Chapter 5 presents the findings of the individual case study analysis of the four Indian-German urban climate partnership projects (the Decentralised Waste Water

1.4 Organisation of the Study

Treatment (DEWATS) and tramway projects as part of the Pune-Bremen partnership; the Waste-to-Energy (W2E) project involving Nashik and Hamburg; and the Local Renewables Model Communities Network (LRMCN) cooperation between Nagpur and Freiburg). Chapter 6 compares the findings of the within-case analysis to test and refine the research hypotheses on key conditions for success and failure in urban partnerships, by identifying cross-case patterns and interlinkages between the explanatory variables and the project outcomes. The broader validity of the key challenges and dilemmas that the four transnational urban partnerships face is discussed in chapter 7, which also outlines the implications of the study's findings for theory and methodology development.

2 State of Research and Key Debates

2.1 Urban Climate Governance

Research on urban climate governance began in the mid-1990s, parallel to the introduction of the UNFCCC Kyoto Protocol (Betsill and Bulkeley, 2007). Over the last two decades the scholarly debate on cities and climate change has centred on several key areas.

One strand of research has explored the policy areas of urban climate action and how the scope of urban climate activities has been enabled and constrained by the legal competencies of cities in climate-related policy fields. Betsill and Bulkeley (2004, 477) point out that "local governments will be critical players in any attempt to implement national and international policy imperatives to reduce emissions of greenhouse gases, and have a significant role to play in climate protection in their own right." They highlight that energy and transport management and urban planning are areas in which most local authorities can make significant contributions to the reduction of GHG emissions (ibid.). Alber und Kern (2008) confirm that most local governments have sufficient legal responsibilities in the areas of energy, transport, urban planning and land-use to be able set up urban climate change programs. They find that although cities also usually control waste management, this is rarely included in local climate strategies.

The capacities and means of urban climate governance have also been comprehensively researched (amongst others by Alber and Kern, 2008; Bulkeley and Betsill, 2013; Bulkeley and Kern, 2006; Schroeder and Bulkeley, 2009). Referring to Bulkeley and Kern (2006), Alber and Kern (2008) distinguish between four modes of urban climate governance. First, 'self-governing'; this encompasses all areas where local governments act as consumers, e.g. public procurement and the energy-efficient refurbishment of buildings owned by the municipality. Alber and Kern argue that climate protection measures in this area can be helpful for agenda setting and the demonstration of political leadership, but the actual impact on reductions in GHG emissions appears to be rather limited. Self-governance therefore needs to be complemented with activities in other governance modes (ibid.). Second, 'governing through enabling'; this includes all activities through which municipalities promote and facilitate voluntary action by local citizens and businesses, e.g. by incentivising renewable energy installations and conducting energy efficiency campaigns. Third, 'governing by provision'; municipalities can also engage in climate protection activities in their role as a service provider of public energy, transport and waste management, e.g. by improving the fuel efficiency in the public transport fleet and by fostering the reuse and recycling of municipal waste. Fourth, 'governing by authority'; many municipalities are also able to enforce local climate protection measures via their mandate as a regulator, e.g. by introducing energy-efficient building standards or local speed limits for vehicles. Alber and Kern however find that municipalities are often reluctant to apply such command-and-control measures as they fear local resistance from political opponents, citizens and businesses (ibid.). Bulkeley and Betsill (2013) argue that in addition to municipal voluntarism (under which the four modes of urban climate governance can be subsumed) cities are increasingly widening their focus and trying to influence and shape national and supranational climate change agendas. According to the authors, this movement of strategic networking and intervention started in 2005 with the United States (U.S.) Conference of Mayors Climate Protection Agreement[10] which was replicated in Europe with the formation of the Covenant of Mayors in 2008 (see section 1.1.2). Moreover, Betsill and Bulkeley find that cities increasingly involve private actors in the design of climate-friendly and resilient urban infrastructure initiatives (ibid.).

A third, related research area has been the role of leadership in urban climate governance. Several studies highlight that leadership by engaged local individuals (who are often termed "policy entrepreneurs") is a crucial condition for the successful setting up and implementation of local energy and climate strategies (Beermann, 2009; Campbell, 2012; Schreurs, 2008). The importance of leadership has also been discussed with regard to the role of cities as laboratories for experimentation within the multi-level governance of climate change (Acuto, 2013; Anguelovski and Carmin, 2011; Cástan Broto and Bulkeley, 2013; Hodson and Marvin, 2010; Jänicke, 2013; Schreurs, 2010). Anguelovski and Carmin (2011) argue that the multiplicity of urban climate experiments worldwide challenge the traditional perspective that local climate action is induced in a top-down manner by external actors (e.g. national governments, donors, non-governmental organisations or TMNs). They find that most local initiatives are in fact rather independent and motivated by internal goals. Jänicke (2013, 13) confirms that the local level has turned into "*the most dynamic level of technical change* towards a low-carbon energy system"[11]. However, Jänicke questions whether local climate action can

---

10   The U.S. Conference of Mayors Climate Protection Agreement is a network of currently 1,060 U.S. cities that have committed to a local implementation of the Kyoto GHG emissions reduction targets in the absence of an ambitious national climate policy (see http://www.usmayors.org/climateprotection/agreement.htm (19-02-2016)).
11   Accentuation by the author in the original text.

really be driven independently of higher policy levels, pointing out that national governments and the European Union are still instrumental in leading and fostering low carbon innovation. Also the role of eco and low carbon model cities is a controversial discussion topic. Schreurs (2010, 97) highlights the potential of environmental model cities in Japan and China to serve as "test beds for new ideas for urban transformation toward low carbon societies" and "models for other cities to follow". This view is contested by Hodson and Marvin (2010) who criticise the emerging concept of eco model cities, arguing that these are often designed in a socially excluding manner and should therefore not serve as models for replication.

A major shortcoming of urban climate governance research remains the narrow focus on individual case studies on large cities from industrialised countries. Comparative perspectives and studies about climate action in small and medium-sized cities and cities from the Global South are scarce. In particular the demand for more comparative research including cities from developing countries has been repeatedly voiced by leading urban climate governance authors (Alber and Kern, 2008; Betsill and Bulkeley, 2007; Castán Broto and Bulkeley, 2013; Rosenzweig et al., 2010). Betsill and Bulkeley (2007) consider the lack of research on cities from the Global South "somewhat surprising" (ibid., 453), as cities from developing countries are becoming increasingly active in local climate responses. Referring to the few studies on urban climate action in Southern cities that were available at the time of their analysis (such as Dhakal, 2004, 2006; Holgate, 2007; Romero-Lankao, 2007), Betsill and Bulkeley identify that cities from the Global North and Global South face similar challenges resulting from climate change, such as the lack of human and financial capacity and political competition with other local issues. Acknowledging that these challenges are often more pressing in Southern cities, Betsill and Bulkeley argue that there is enough common ground for more international comparative research on urban climate governance (Betsill and Bulkeley, 2007). In their 2013 update of urban climate research Bulkeley and Betsill (2013) review additional research contributions on urban climate response in cities of the Global South since 2007 (such as Aylett 2011; Bulkeley et al. 2009; Hardoy and Romero Lankao 2011; Kithiia 2011). They conclude that these studies support the trend that the number of Southern cities engaged in the development of climate policy is continuing to grow, largely due to the revitalisation of transnational climate networks and the increasing focus on public-private partnerships in urban climate responses. However, most studies on urban climate responses still show a geographical bias towards cities from industrialised countries in North America, Australia and Europe, as Castán Broto and Bulkeley (2013) find in their survey of local climate experimentation in 100 cities around the world. Rosenzweig et al. (2010, 911) reach the similar conclusion specifying that particularly smaller cities' climate activities have not received sufficient attention from the research community: "research networks need to be expanded to include more cities across both the developed and the developing worlds — especially small or medium-sized cities, in which limited resources need to be utilized as efficiently as possible."

A second limitation of urban climate governance research has been the lack of literature on urban climate action conducted by non-state actors. Castán Broto and Bulkeley (2013) address this research gap in one of the first major studies on the state of urban climate governance worldwide. Their survey reveals surprising results with regard to the importance of private actors in local climate experimentation indicating that globally non-state actors account for about one third (34%) of local climate action. The survey also uncovers a remarkable regional feature, namely that in Asia almost half (47%) of urban climate initiatives are driven by non-state actors. Castán Broto and Bulkeley therefore urge that the research focus on urban climate governance is to be widened to include non-state actors and climate governance outside formal policy channels, which according to their findings have become key players in climate action at the city level.

2.2 Urban Climate Governance in Germany and India

The following section introduces on-going developments, specific features and key differences in urban climate and low carbon governance in German and Indian cities, the focus of this study's analysis.

German and Indian cities' experiences in urban climate policy-making vary substantially. In Germany, frontrunner cities have been setting up local climate protection strategies for more than 25 years. Today most large and medium-sized cities have adopted climate policies and even many smaller towns and municipalities have introduced action plans to reduce their local carbon foot-prints. Many German cities are implementing comprehensive policies covering multiple climate change-related sectors such as renewable energy development, public transport, urban planning, land-use and others. Several German cities, for example Freiburg, Hannover and Münster, have established climate change departments within their city administrations (Deutsches Institut für Urbanistik, 2011). To improve their national and international reputation, a growing number of cities promote their strengths by branding themselves via labels such as "Wind Energy Capital" (Hamburg), "Solar City" (Gelsenkirchen), "Bicycle Capital" (Münster) or "Green City" (Freiburg).

Indian cities started experimenting with climate and low carbon policies much more recently. In 2007, Nagpur and Bhubaneswar were among the first Indian cities to introduce city-level low carbon policies, adopting renewable energy and energy efficiency policies (see section 5.4 of this study). In 2009 Delhi was the first Indian city to launch a comprehensive climate action plan, the Delhi Climate Change Agenda 2009-2012, covering 65 fields of action to be conducted by all departments of the Delhi government (Sharma and Tomar, 2010). Over the following years, several additional Indian cities, amongst others Kolkata (2010), Rajkot and Coimbatore (2011) introduced climate action programmes, many in cooperation with the transnational municipal network ICLEI (ICLEI South Asia, 2011). In comparison to the comprehensive climate policies of many German cities, Indian cities' climate and low carbon initiatives are often less integrated and built around a limited number of projects in a certain policy field (Sharma and Tomar, 2010).

Three reasons explain the later and less comprehensive introduction of low carbon policies in Indian cities. The first determining factor is the differing degrees of legal responsibilities in German and Indian cities with regard to cli-mate and low carbon policy making. In Germany's federal political system, the constitution ("Grundgesetz") grants municipalities the right to self-government, including the fiscal responsibility, for all local matters (Grundgesetz Article 28(2)). Climate policy is a task that is carried out voluntarily by municipal self-government, which means that cities decide themselves whether and how they pursue climate strategies (Deutscher Städtetag, 2010). Thus, most cities design their climate policies largely independently of subordinate policy levels such as state and central governments (Deutsches Institut für Urbanistik, 2011; Hakelberg 2011). Hertle and Schächtele (2008, 4) confirm that German cities are not constrained by major legal restrictions to climate policy-making:

> Many German communities support climate protection measures without being tied to a strict separation of tasks on national, regional and local level. Communities may introduce administrative regulations (i.e. energy standards), financial incentives and soft instruments to push local climate protection. (...) Local governments' possibilities to influence climate protection are manifold and resemble the policies in Germany.

Ohlhorst, Tews and Schreurs (2013) point out that Germany's federal political system generally facilitates social and institutional innovation at the sub-national level. They explain that due to the largely decentralised nature of renewable energy technologies, they are often promoted by the local level. In addition to renewable energy development many German cities' climate strategies focus on improving energy efficiency in the building sector, and fostering low-carbon mobility such as public transportation, cycling and walking. To set an example for local citizens and businesses city governments often target policy areas over which they have direct influence, such as introducing emission-free municipal car fleets, refurbishing municipal buildings and promoting renewable energies in municipal public utilities. Moreover many local governments offer energy consulting for households, conduct awareness-raising projects and involve citizens in the development and implementation of climate strategies (Climate Alliance - Klima-Bündnis, 2008; Deutsches Institut für Urbanistik, 2011; Städtetag, 2010).

Whereas German cities set up climate strategies largely independently, Indian cities rely much more on central and state governments in their development of local climate action. Climate governance in India is generally designed in a top-down manner and the central government ministries and states are the dominant players in the development of India's climate policy. Municipal bodies help implement these policies, but they are bound by clear guidelines set by state and central governments (Sharma and Tomar, 2010). The scope for self-government is limited, as Indian cities lack decision-making competencies in key climate-related policy fields such as transport and energy production and distribution. In their study on climate change and urbanisation in India, Mukhopadhyay and Revi (2012) explain that according to the 74th Amendment to the Indian constitution of 1992, cities should be given greater autonomy over local policy matters such as town planning. However, the authors conclude that the amendment has been poorly implemented and most states have failed to effectively transfer town planning competencies to cities. Therefore, Indian cities have to cooperate closely with state and central governments if they wish to develop and implement comprehensive local climate and low carbon policies.

A second, related reason for the later introduction of urban low carbon and cli-mate policies in India is the lack of financial means. The problem of limited funds for climate action is already pressing in most German cities, but it is even more severe in Indian cities.

Despite having more legal and fiscal competencies than Indian cities most German cities still strongly rely on financial support from higher policy levels. In fact, municipalities in Germany face the challenge of seeing their re-sponsibilities increase while their financial means decrease because of the recent economic crisis and rising social expenditures (Meyer-Timpe, 2010). Thus, many German cities have to deal with budgetary deficits and have been forced to reduce the voluntary tasks associated with municipal self-government, such as for example climate policy. Programmes fostering investment in the energy-efficient refurbishment of buildings, energy consulting for citizens or the establishment of the post of a municipal climate protection officer are often withdrawn. Germany's central government and the European Union have reacted to municipalities' financial problems by setting up a number of funding programmes which support cities and towns develop and implement climate protection strategies, e.g. the German government's National Climate Protection Initiative (NKI) or European programmes such as the Intelligent Energy-Europe (IEE), Life+, Smart Cities and Communities Initiative and European Energy Efficiency Fund (see Climate Alliance-Klima-Bündnis, 2012 for a detailed overview). Several cities facing severe budgetary crises have had to establish "emergency budgets" which reduce their scope of action to such an extent that they are not even able to apply for loans or support schemes from the German government or the European Union (Deutsches Institut für Urbanistik, 2011).

Compared to their Indian counterparts, German cities still have better access to funds for climate and low carbon projects. Bhagat (2005, 68) explains that Indian cities are relatively inexperienced when it comes to budgetary planning and responsibility as until the mid-2000s, state governments used a "gap-filling approach" to financially support cities. Bhagat points out that cities were only recently expected to generate their own funds, mainly through property and vehicle tax revenues, some non-tax revenues such as rents from municipal assets, plus external sources, such as grants (ibid.). Local governments in India however struggle to collect tax revenues efficiently, so that they often lack the financial resources to deliver even basic services in the most pressing policy are-as such as energy, water and waste management (Sharma and Tomar, 2010). Climate mitigation and low carbon development are of only secondary or even no concern at all to many local decision makers and are usually only introduced when they offer clear financial co-benefits or when they are fully funded by external schemes (Bhagat, 2005; Mukhopadhyay and Revi, 2012).

A third reason why Indian cities' engagement in climate action lags behind their German counterparts is the widely-shared perception in India that the industrialised countries such as the United States, European countries and Japan caused global climate change and therefore should be held accountable for its repercussions (Agarwal and Narain, 1991; Dubash 2012a, 2012b). Fisher (2012, 109) finds that Indian climate politics is still "largely dominated by the state position in international negotiations" according to which responsibility for climate protection lies primarily with industrialised countries, whereas India as a developing country needs to prioritise poverty alleviation and economic development and therefore cannot accept any binding GHG emissions reduction targets. However, Fisher outlines that non-state and subnational actors increasingly shape "aspects of the debate" and calls for more research on local politics to enable "a more rooted understanding of Indian climate politics." (ibid.) Dubash (2012a) confirms that India's international focus is on equity and the responsibility of wealthy countries. At the same time he recognises growing concerns within India that climate change will adversely affect India's economic development. The primary focus of Indian climate policy – both at the national and city level – is therefore on adaptation and disaster management (Mukhopadhyay and Revi, 2012; Nair, 2009; Revi, 2008). Mitigation without co-benefits remains a niche topic in India, despite the introduction of the National Action Plan on Climate Change (NAPCC) in 2008 and India's approval of the Paris Agreement in December 2015.

2.3 Urban North-South Cooperation

Urban exchange and learning between cities from the Global North and Global South has been gaining popularity as a research topic since the early 2000s. Existing studies distinguish between three forms of transnational urban cooperation, based on dominant actor structures: state actor-driven cooperation, private actor-driven cooperation and cooperation driven by transnational municipal networks.

2.3.1 State Actor-driven Transnational Urban Cooperation

State actor-driven urban North-South cooperation has been subsumed under the terms of "decentralised cooperation", "city-to-city cooperation" and "municipal international cooperation".

Hafteck (2003) explains that the concept of "decentralised cooperation" (DC) was developed in the 1980s by governmental institutions in Europe, North America and Japan to address the emergence of cities as new actors in development cooperation activities. He points out that national governments started to involve cities more closely in their development aid strategies as a response to the apparent challenges of urbanisation and the increasing relevance of the principle of subsidiary and social issues in development aid. Unlike NGOs, which at that time were under pressure to prove their efficiency in conducting development aid, local governments were seen as qualified partners for development projects, as institutions with in-house technical, financial and planning expertise and often already with established international relations, such as twinning partnerships. Many countries, for example Canada (1987), Italy (1987), Japan (1988), France (1992) and the UK (1993) formally institutionalised the concept of DC into laws and programs. Hafteck identifies three commonalities which the diverse interpretations of DC shared. Most definitions consider local governments as the lead actors of DC and emphasise the need for formal institutionalisation of DC in written agreements between the local governments of the cities involved. Furthermore, the majority of DC

2.3 Urban North-South Cooperation

definitions highlight sustainable local development as the major target of DC and exchange and support as its main activities (ibid.).

From around 2000 the concept of DC was gradually replaced by the concept of "city-to-city cooperation (C2C)", in both public as well as academic discourse (ibid.). Bontenbal and van Lindert (2009) reveal that the term "city-to-city cooperation" was introduced by the United Nations Environment Program (UNEP) in 2000 and popularised UN-Habitat in 2002 when it was chosen as the theme for the World Habitat Day. Apart from a focus on local governments as major actors in urban cooperation, there is, however, no widely agreed upon definition of C2C, as Bontenbal and van Lindert point out (ibid.). Another term often used to describe state-led partnerships between cities from the Global North and Global South is "municipal international cooperation" which van Ewijk and Baud (2009) suggest is a more generic term compared to C2C which according to them is limited to North–South cooperation between smaller municipalities.

Despite the lack of a widely-recognised definition, existing studies still offer insight into recent developments in state actor-led urban cooperation. Van der Pluijm and Melissen (2007) identify a shift towards more professionalisation and pragmatism in city-to-city exchange. They highlight that cities have been setting up international relations and networks since ancient Greek times and that at the beginning of the 21st century cities are once again getting involved in international diplomacy, fostering international cooperation and influencing international organisations. Van der Pluijm and Melissen conclude that urban partnerships increasingly focus on concrete project development and economic growth rather than following idealistic motives (ibid.). In their analysis of relations between Dutch cities and cities from the Global South, van Ewijk and Baud (2009) confirm this trend towards more pragmatism and project-orientation in urban cooperation. They find that because of globalisation, local governments are impacted more and more by events happening outside of their city's borders and therefore need to develop new governance approaches. Van Ewijk and Baud identify two simultaneous trends in urban cooperation in the Netherlands. Dutch cities increasingly pursue mutuality in urban partnerships, setting up relations with countries of outward migration to the Netherlands, such as Morocco, Surinam and Turkey with the aim of improving the integration of migrant communities and fostering their own economic development. At the same time, cities continue to focus on international solidarity and sustainable development in their North-South partnership work, supporting the targets of the United Nations Millennium Development Goals (ibid.). Campbell (2012) points out that most cities still do not do enough to utilise opportunities to exchange knowledge and experiences. He argues that many cities are energetic, but disorganized; productive, but still not efficient; promising, but lacking channels to reach application more widely where they are needed. Above all, the barriers of institutions, distant policy and isolated practice can be cleared away by activating one of the most potent but underutilized ressources available to address urban problems: knowledge already invented in or by other cities. (ibid., 204).

Only a few studies have looked more in-depth at the conditions for success and failure in local governments' transnational urban cooperation and exchange activities. Tjandradewi and Marcotullio (2009) address the research gap on how city-to-city cooperation "actually works" (ibid., 166), analysing Asian city managers' perspectives on success conditions for urban cooperation. Their survey confirms the relevance of four factors in particular; free flows of information, reciprocity, mutual understanding, and leadership. Tjandradewi and Marcotullio were surprised to find that city officials did not consider community participation as a highly relevant aspect for successful city partnerships (see also section on "Private Actor-driven Transnational Urban Cooperation" below).

Bontenbal and van Lindert (2009) point to persisting research gaps in the area of state-led urban cooperation. They highlight that research has not yet acknowledged the increasing importance of urban cooperation in global North-South relations: "Although the number of C2C arrangements, city networks and local authorities involved in international cooperation is substantial, C2C is a fairly recent theme in the academic debate on development cooperation." (ibid., 131). Bontenbal and van Lindert conclude that research on C2C is generally limited and fragmented, and that research gaps remain in the areas of "objectives and results, organisational structures, success factors and weaknesses" (ibid.).

2.3.2 Private Actor-driven Transnational Urban Cooperation

Research is also scarce regarding the role private actors play in driving and participating in transnational urban cooperation. Bontenbal and van Lindert (2008) find that C2C theoretically offers great potential to bridge local governments and civil society and thereby "touch upon the core of urban governance" (ibid., 479). But in practice urban North-South partnerships often struggle to involve and mediate between state and civil society actors. Bontenbal and van Lindert conclude that C2C tends to have a greater impact on improving municipal institutional performance than on civil society empowerment. They also identify a general problem associated with civil society-driven partnerships is that they often remain ad hoc and act in isolation to other governance processes within cities (ibid.).

Tjandradevi and Marcotullio (2009) confirm that a pertinent gap remains between requests for more civil society involvement and its practical implementation in urban cooperation. They find that while the United Nations Development Programme (2000) lists community participation as one of five key success conditions for setting up C2C, city officials involved in urban partnerships rank civil society participation as the least important out of nine success factors offered in Tjandradevi and Marcotullio's survey.

2.3.3 Urban Cooperation in Transnational Municipal Networks

A third form of urban cooperation which has gained increasing attention from the research community is collaboration facilitated by TMNs. In particular ICLEI and its Cities for Climate Protection (CCP) campaign have been the focus of several studies. Betsill and Bulkeley (2006, 141) argue that the CCP exemplifies a new form of governance in global efforts to mitigate climate change, being "simultaneously global and local, state and non-state" and taking place "through processes and institutions operating at and between a variety of scales and involving a range of actors with different levels and forms of authority".

An assessment of the role and impact of TMNs is however mixed. Alber and Kern (2008) list cities' involvement in TMNs as one of four enabling factors for local climate policy development (in addition to the (perceived) climate change impact, cities' competencies and authority to regulate climate change and national government support schemes). Bulkeley and Newell (2010, 59) also highlight TMNs as "one of the first and most extensive examples of transnational governance". They explain that there have been two phases of TMN development, with the first peak in the early 1990s leading to the establishment of TMNs such as ICLEI/CCP, the Climate Alliance and Energy Cities in Europe and North America. Since the mid-2000s a second wave of TMN growth has been driven by globally-oriented TMNs such as C40 and the World Mayors Council on Climate Change. Bulkeley and Newell identify a "more avowedly political nature of this second wave", and point out that TMNs exert increasing influence on the international level (ibid., 60). The European Covenant of Mayors, another more recently established TMN, has been praised for its ability to foster local renewable energy and climate action by addressing economic interests and facilitating progressive competition among local governments through benchmarking mechanisms (Jänicke, 2013).

While the growing membership and international visibility of TMNs is undisputed, their ability to foster knowledge exchange among its members has, however, been repeatedly questioned (Betsill and Bulkeley, 2004; Medearis and Dolowitz, 2013). Medearis and Dolowitz (2013) state that TMNs such as ICLEI, Metropolis and the C40 deserve merit for providing information about best practice examples of urban sustainable innovations. But they criticise TMNs for not having sufficiently addressed the complexity of policy transfer processes and not having been able to facilitate "problem-focused, goal-oriented, resource-relevant, and geographically-specific exchanges of urban development policies" (ibid., 233). TMNs are also criticised for being "networks of pioneers for pioneers" and having too many passive members (Kern and Bulkeley, 2009, 311). Kern and Bulkeley conduct a comparative analysis of German and British cities' participation in the three TMNs ICLEI, Climate Alliance and Energy-Cities. They find "fundamental differences between active and passive network members" (ibid.) and that TMNs "reinforce differing patterns of network participation between leaders and laggards" (ibid., 328-329).

As these studies demonstrate, most research on TMNs is from the perspective of the TMN itself. Research focusing on cities and their motivation to join TMNs, processes of city collaboration in TMNs and the cities' (perceived) benefits and shortcomings of TMN membership is rare. An exception is Hakelberg (2011) who analyses two German cities' (Hanover and Offenbach) engagement in TMNs and finds that "TMN membership is indeed the prime motivator for a city's adoption of a local climate strategy, mainly because networks succeed in facilitating learning processes among their members." (ibid., 3). A second exception is Nakamura (2010) who studies the endogenous and exogenous political factors in Indonesian cities that led to the adoption of new environmental policies via ICLEI's CCP initiative. Nakamura identifies four success conditions for an effective participation of cities in the CCP: political leadership by the local mayor, the activation of local stakeholders within the local government and beyond, sufficient political and financial autonomy of local governments, and support from officials at higher policy levels.

2.4 Transnational Urban Cooperation in Germany and India

The following section presents an overview of existing knowledge and academic literature about the particularities of transnational urban relations in German and Indian cities, with a special focus on climate and low carbon related exchange activities.

2.4.1 Transnational Urban Cooperation in Germany

Many German municipalities have established transnational urban relations. A database by the Rat der Gemeinden und Regionen Europas (RGRE) shows that German municipalities have set up a total of 5434 formal "partnerships", 610 project-oriented and temporary "friendships" and 1079 informal "contacts" with municipalities outside of Germany"[12].

There are however few partnerships between German cities and cities from the Global South. Nitschke, Held and Wilhelmy (2009), in reference to Heinz and Leitermann (2004)) state that development cooperation only features in about 3% of all formal city partnerships of German cities. According to their analysis German municipalities are more active in fostering development cooperation through the promotion of fair trade, fair procurement and development education; but they are less active in establishing partnerships with cities from the Global South. Nitschke et al. (ibid., 135) believe that "the potential of German cities for cooperative development projects is not yet fully realised". They outline several challenges facing German cities in establishing partnership projects with cities from the Global South, such as having lower levels of institutional and financial support for development activities than cities in other European countries (ibid.). Despite German cities having relatively strong decision-making powers in climate-related policy fields (see section 2.2 of this study) in the establishment of international partnerships their competencies are weaker. In Germany's federal political system, the mandate for international relations lies primarily with the central government. Thus, German cities are under more pressure to demonstrate that their development cooperation activities are in line with the central government's international development cooperation framework and to prove the quality and efficiency of their development cooperation activities. In the absence of adequate institutional and financial support many smaller and medium-sized city governments abstain from engaging in development cooperation (ibid.). In their analysis of municipal sustainability partnerships and networks Statz and Wohlfahrt (2010) confirm that it is primarily larger German cities that have the organisational and financial capacity to maintain active international relations with partner cities and multilateral networks. They highlight that the city states of Berlin, Hamburg and Bremen are able to utilise their broader legal mandates and more extensive financial resources to conduct urban development and economic cooperation. In most small and medium-sized German cities urban development partnerships are often driven and maintained by civil society actors such as partnership associations, often with rather limited budgets (ibid.; Nitschke et al., 2009).

---

12   http://www.rgre.de/partnerschaften0.html (database query from 06-03-2015).

Sustainable development and climate change are key areas of cooperation for German cities, especially in partnerships with cities from the global South (Statz et al., 2010). The city of Bremen, for example, has been working on environmental, social and low carbon projects with international partner cities since the 1970s, largely conducted by non-state individuals and organisations (see sections 5.1-5.3 of this study). Other municipalities in which environmental and climate change projects feature in their partnership activities include Lauingen (cooperation with Lagos Island, Nigeria), Ludwigshafen (exchange with Sumgait, Aserbaidschan), Dresden (partnerships with Lwiw, Ukraine and Wroclaw, Poland) and Erfurt (cooperation with Vilnus, Lithuania and other cities) (Statz et al., 2010).

More recently there have also been efforts at the national governmental level in Germany to pursue urban North-South climate change cooperation. The "50 Municipal Climate Partnerships until 2015" program led by the German Federal Ministry for Economic Cooperation and Development (BMZ) has helped selected German cities collaborate with African and Latin American cities on matters related to climate change and low carbon development (ENGAGEMENT GLOBAL – Service for Development Initiatives, Service Agency Communities in One World, 2014).

During the last two decades, many German cities have also engaged in TMNs, particularly those pursuing sustainability and climate change policies. The TMN with the largest number of German member cities is the Climate Alliance (480 German member cities)[13], followed by other popular TMNs, such as the European Covenant of Mayors (56)[14], ICLEI (17)[15], Energy Cities (8)[16], METRIX – The Network of European Metropolitan Regions and Areas (8)[17], the Carbonn Climate Registry (3)[18], C40 (2)[19], Metropolis (1)[20] and the World Mayors Council on Climate Change (1)[21]

---

13 http://www.klimabuendnis.org/fileadmin/inhalte/dokumente/2015/Mitgliederliste_Deutschland_Februar_2015.pdf (19-03-2015)
14 http://www.eumayors.eu/covenant_signatories.pdf (19-03-2015)
15 http://www.iclei.org/iclei-members/iclei-members.html?memberlistABC=I (19-03-2015)
16 http://www.energy-cities.eu/cities/members_in_europe_en.php (19-03-2015)
17 http://www.eurometrex.org/ENT1/EN/Members/members.php (19-03-2015)
18 http://carbonn.org/data (19-03-2015)
19 http://www.c40.org/cities (19-03-2015)
20 http://www.metropolis.org/map?field_cities_region_value=348 (19-03-2015)
21 http://www.worldmayorscouncil.org/members/members-list.html (19-03-2015)

2.4.2 Transnational Urban Cooperation in India

City partnerships focusing on sustainable and low carbon development are also driven increasingly by Indian cities. Several Indian cities have partnered with cities from the Global North to focus on issues other than traditional twinning activities such as cultural and individual citizen exchange. These partnerships are often facilitated and moderated by third parties. For example, Delhi's collaboration with Tokyo on the development of a new metro system for the Indian city is facilitated by the Japanese International Cooperation Agency (JICA). Ahmedabad's partnership with the Spanish city of Valladolid and the German city of Halle (Saale) to develop a comprehensive program of ecological heritage preservation is supported by the NGO EuroIndia Centre. While Guntur (India), Bologna (Italy) and Vaxjö (Sweden)'s cooperation in the introduction of ecoBUDGET, a city-level environmental management system, has been developed by the TMN ICLEI. Indian cities have also partnered with other cities in the South, such as Coimbatore's knowledge-exchanges with Ekurhuleni (South Africa) and Yogyakarta (Indonesia) on renewable energy and energy efficiency strategies as part of ICLEI's "Local Renewables Model Communities Network".

India's increasingly overburdened urban energy, water and transport infrastructure is an important driver for Indian cities to turn to international cooperation as a way of developing innovative and sustainable solutions to these challenges. As urbanisation and industrialisation have grown, so have energy and water consumption, solid waste and sewage streams and the demand for individual and public transport. A study by the McKinsey Global Institute (2010) estimates that in order to meet the demands of an increasing urban population India needs to build 2.5 billion square metres of roads and 7,400 kilometers of metro and subway lines in the transport sector alone by 2030, which is about 20 times the total capacity that was added between 2000 and 2010. Mukhopadhyay and Revi (2012) argue that in view of these challenges Indian cities are more open towards learning from other cities compared for example to their European counterparts. The cities' lack of access to documentation on their peers' experiences in sustainable and climate governance can however hinder learning (Sharma and Tomar, 2010).

Indian cities' reliance on funding and approval from national and state governments is another barrier to them being able to enter formal city partnerships or international partnership projects. Tjandradewi and Marcotullio (2009) point out that Indian cities share this challenge with cities in most other Asian countries, where support from higher political levels is a prerequisite for establishing city partnerships. Even if partnership actors have access to central and state government institutions and receive approval for joint projects, this process often leads to considerable time delays which places strain on the projects' budgets (Beermann, 2014). According to Fisher (2012) the dominance of the Government of India has a greater influence on climate governance than any transnational linkages and the national government's decisions shape climate action at all policy levels including the local level. The scope of action for transnational climate governance networks is therefore limited and these networks often choose to follow the national government's policies rather than oppose them (ibid.).

Interestingly, despite these institutional barriers, India is still considered as a global frontrunner in fostering horizontal exchange at the city level. The Indian government in fact actively promotes learning among cities. In 2007, it introduced the Peer Experience and Reflected Learning (PEARL) program as part of the Jawarhalal Nehru National Urban Reform Mission (JnNURM) to foster city exchange in the areas of solid waste management, water supply and sanitation, urban transport and cultural heritage. 167 major Indian cities were divided into six groups (Mega Cities, Industrial Cities, Mixed Economy Cities, Cultural Cities, Cities of Environmental Importance and North East Cities), to facilitate partnerships between cities with similar socio-economic profiles and interests. Knowledge exchange takes place online via the PEARL website. So far, 37 good practise projects and initiatives have been uploaded by the National Institute of Urban Affairs (NIUA) that coordinates the program.[22] Campbell (2012, 209) states in his study of urban learning worldwide that this program is rather unique: "Only a handful of nations have focused on horizontal exchange as a matter of policy. India is a bellwether."

Whereas the JnNURM's PEARL program focuses on fostering domestic city exchange the body has also provided financial support for local capacity building via international city cooperation. The JnNURM funds have, however, not been widely accessed by city managers, many of whom were apparently not aware that they could apply for them, as pointed out by Rakesh Ranjan, advisor at the Government of India's Planning Commission.[23]

Like German cities, a large number of Indian cities are also engaged in sustainability and low-carbon related TMNs. Several TMNs operating globally have in fact more Indian than German member cities, for example ICLEI (46 Indian member cities)[24], the Carbonn Climate Registry (19)[25], the World Mayors Council

---

22 http://pearl.niua.org/ (19-03-2016)
23 Presentation by Rakesh Ranjan at the 6th EuroIndia Summit on "Greening Cities", 21-22 October, 2013 in Hyderabad.
24 http://www.iclei.org/iclei-members/iclei-members.html?memberlistABC=I (19-03-2015)
25 http://carbonn.org/data (19-03-2015)

on Climate Change (6)[26], Metropolis (6)[27] and the C40 (3)[28]. This is remarkable since compared to German cities, Indian cities have only started engaging with issues of sustainability and climate change more recently, even if of course India's total population and number of cities far exceeds that of Germany and TMN membership alone says little about actual activity (see above). In addition to membership of global TMNs, several Indian cities are also involved in TMNs with a regional focus on Asia. Examples are CITYNET (3 Indian member cities) [29], the Clean Air Initiative (3)[30] and the Kitakyushu Initiative for a Clean Environment (2)[31].

2.4.3 Urban Cooperation between German and Indian Cities

Considering institutional barriers and city managers' lack of knowledge about financial support mechanisms for international collaboration, it is not surprising that the number of partnerships between German and Indian cities is still modest. The above mentioned RGRE database lists six initiatives involving German and Indian cities; three city "partnerships" (Bremen-Pune, Stuttgart-Mumbai, and Herrsching-Chatra), one temporary and project-oriented city "friendship" (Werne-Rourkela) and two informal "contacts" (Esslingen-Coimbatore and Langenhagen-Alan Kuppam).[32] Four of the initiatives have a clear focus on sustainable and low carbon development (Bremen-Pune, Herrsching-Chatra, Werne-Rourkela and Esslingen-Coimbatore, with the latter cooperation still being in the preparatory stage of developing a partnership MoU). The partnership between Stuttgart and Mumbai primarily engages in cultural exchange activities while information available on the cooperation between Langenhagen and Alan Kuppam is very limited.

The RGRE database output should however be treated with caution as the database is not comprehensive and does not list several new forms of city cooperation that have emerged in recent years. In particular, several city partnerships facilitated by TMNs, development agencies or NGOs have so far not been included in the database. One example is the cooperation between Ahmedabad and Halle on ecological heritage conservation, facilitated by the NGO EuroIndia Centre. Two other examples are analysed in this study: the collaboration between Nashik and Hamburg, facilitated by the German government's development

---

26 http://www.worldmayorscouncil.org/members/members-list.html (19-03-2015)
27 http://www.metropolis.org/map?field_cities_region_value=348 (19-03-2015)
28 http://www.c40.org/cities (19-03-2015)
29 http://citynet-ap.org/citynet-members/ (19-03-2015)
30 http://cleanairinitiative.org/portal/countrynetworks/india?page=2 (19-03-2015)
31 http://kitakyushu.iges.or.jp/cities/index.html (19-03-2015)
32 http://www.rgre.de/partnerschaften0.html (database query from 06-03-2015).

agency, the GIZ, and the partnership between Nagpur and Freiburg which was initiated and moderated by the TMN ICLEI (see sections 5.4-5.7 of this study). Aside from the projects mentioned above collaboration between German and Indian cities is rare. The supervisor of the GIZ-facilitated cooperation between Nashik and Hamburg's public water utility Hamburg Wasser, recalls two other GIZ initiatives involving Hamburg and Indian cities: an art project on river water management fostering exchange between artists from Hamburg and Delhi in 2011/2012 and a recent program in which students from Varanasi travelled to Hamburg to exchange experiences with young Germans taking part in the "Voluntary Ecological Year" program" (Interview with GIZ, 16-12-2012). As far as the author is aware these are the only Indian-German city cooperation projects being conducted in the areas of sustainable or low carbon development. According to representatives of ICLEI's regional offices in Europe and South Asia, the collaboration between Nagpur and Freiburg is the only Indian-German city partnership facilitated by ICLEI (Interviews with ICLEI, 08-08-2012 and 05-12-2013). The interviewee from the TMN Climate Alliance is not aware of any cooperation between Indian and German cities as part of Climate Alliance's activities (Interview with Climate Alliance, 19-01-2012). The Ahmedabad-Halle partnership is also the only German-Indian city cooperation to be initiated by the EuroIndia Centre. Secretary General, Michel Sabatier, however states that in the coming years the Centre's focus will be on the establishment of additional city partnerships between European and Indian cities in the area of low carbon development.[33]

So far, research on Indian-German city partnerships is limited. Only the partnership between Pune and Bremen, which is also analysed in this study, has previously been the focus of comprehensive research. Sippel (2007) studied the Pune-Bremen cooperation as one of three cases in her doctoral thesis exploring the capacity of North-South city partnerships to foster projects under the UNFCCC Clean Development Mechanism (CDM). Sippel concludes that although city partnerships offer some potential for CDM project development, at the time of completion of her study (2007) no CDM projects had been set up in any of the partnerships she had analysed. To the author's knowledge no other comparative research on the conditions for success and failure of Indian-German urban partnerships currently exists.

---

33   Presentation by Michel Sabatier at the 6th EuroIndia Summit on "Greening Cities", 21-22 October, 2013 in Hyderabad.

3 Theoretical Framework

To explore the conditions for success and failure in transnational urban partnership projects the study derives four explanatory variables from current theoretical literature. The first explanatory variable, the quality of the knowledge exchange strategy, is derived from policy transfer literature. To position transnational urban cooperation within the wider global climate governance debate, this study engages with emerging transnational governance literature, in particular Andonova, Betsill and Bulkeley's (2009) typology of transnational climate change governance networks. As this typology lacks guidance with regard to the conditions for success and failure of transnational cooperation the study derives additional explanatory variables from New Institutionalism literature, highlighting the role of institutional linkages with the state system; and from Policy Entrepreneur and Social Capital literature analysing the relevance of agency by individuals and collectives in transnational urban partnerships. For each explanatory variable a hypothesis is established to be tested in the comparative case study analysis.

3.1 Policy Transfer Theory

The first explanatory variable is derived from policy transfer research, to investigate how and under which conditions ideas and experiences travel between cities in transnational urban climate cooperation.

Policy transfer theory analyses individual processes of the transfer of knowledge and political programs across distance, usually between two entities. Other than the two related theoretical strands of policy diffusion and policy convergence literature that focus on macro level policy dissemination, policy transfer theory analyses exchange and cooperation processes at the micro level and is often conducted in qualitative case study analyses (Holzinger, Jörgens and Knill, 2007; Mossberger and Wolman, 2003; Wolman and Page, 2002). Dolowitz and Marsh (2000, 5) define policy transfer as a "process by which knowledge about policies, administrative arrangements, institutions and ideas in one political system (past or present) is used in the development of policies, administrative arrangements, institutions and ideas in another political system." Dolowitz und Marsh identify a total of eight possible components of policy transfers, such as policy goals, policy content, policy instruments, policy programs, institutions, ideologies, ideas and attitudes and negative lessons (ibid., 12). Referring to Rose (1993), Dolowitz and Marsh (2000) also create categories of different forms and degrees of policy transfers, ranging from copying, emulation and combinations of different policies to inspiration for policy change.

Several studies offer insight into the conditions leading to success and failure in policy transfers. Tews (2008) identifies the political enforceability of a transferred policy in a new setting as a key factor for success, specifying that political enforceability depends on the capacity of the actors involved to activate political networks and utilise policy windows of opportunity, a factor which is also highlighted by Dolowitz and Medearis (2009). A second condition for successful policy transfers outlined by Tews (2008) is the technical feasibility of transferring a program to a new setting; the conclusion being that it may be advisable to launch transfer projects with pilot initiatives that demonstrate their future technical feasibility.

Dolowitz and Marsh (2000) analyse the causes of unsuccessful policy transfers. They find that most policy transfers are constrained by the "bounded rationality" (ibid., 13) of the actors involved which results from limited or inadequate information about the transferred program and the respective contextual conditions. Dolowitz and Marsh (ibid., 17) explain that transfers often fail when the actors of the borrowing setting lack knowledge about the transferred policy or institution and how it operated in the original setting ("uninformed transfer"); when they have not considered crucial components that made the policy or institution a success in the original setting ("incomplete transfer); or when they neglect the respective contextual conditions in the original and new setting ("inappropriate transfer"). Wolman and Page (2002) and Mossberger and Wolman (2003) point out that for the recipients of a transferred program it is particularly challenging to evaluate the influence of the contextual conditions on the performance of the program in the original context. Mossberger and Wolman (ibid., 437) therefore recommend involving knowledgeable moderators that "can translate between the original and borrowing countries [which] can foster awareness and knowledge of differences in policy settings and the implications of these differences."

Medearis and Dolowitz (2013) confirm that the lack of adequate evaluations of programs and their specific contextual conditions is a key barrier to successful policy transfers. They argue that while global networks such as ICLEI, C40, the Organisation for Economic Development and Cooperation (OECD) and UN Habitat can be praised for collecting and providing information about cases of good practice in urban sustainability innovations, they have not yet sufficiently focused on preparing, implementing and following up on actual policy transfers among their member cities. Medearis and Dolowitz state that a crucial strategy for addressing this challenge is to facilitate longer term personal interaction between city actors on "problem-focused, goal-oriented, resource relevant and geographically-specific exchanges of urban development policies" (ibid., 233).

In this study, propositions on the conditions for success and failure in policy transfers are adapted to the context of transnational urban climate partnerships and tested in case studies according to the following hypothesis:

Hypothesis 1: A transnational urban partnership project is more likely to succeed if it follows a well-prepared knowledge exchange strategy.

The study defines the indicators of a "well-prepared knowledge exchange strategy" as the prior evaluation of the transferred program and of the respective contextual conditions in both partner cities; a systematic approach to the transfer; continuity of interaction between the partnership actors; the involvement of external partnership moderators; the utilisation of policy windows; and the addressing of the intrinsic interests of both partner cities (see Index System in section 4.3 of this study).

The indicator of intrinsic interests has been included to account for the particularities of urban South-North exchange. Several studies show that urban North-South cooperation often struggles due to a (perceived) one-sided flow of learning from cities in the Global North towards cities in the Global South (Bontenbal and van Lindert, 2009; Johnson and Wilson, 2009; van Ewijk and Baud, 2009). These studies argue that in order to realise more equality and mutuality in partnerships, both partner cities need to be able to identify and promote intrinsic interests in the cooperation. The indicator of intrinsic interests speaks to two areas that are as yet underdeveloped in policy transfer literature but which are highly relevant to transnational urban cooperation: the role and motivation of the producers and senders of information and exchange of knowledge and experiences between the Global North and Global South. Wolman and Page (2002, 479) highlight that in order to better understand cross-national policy transfers, literature needs to address the role of actors from the countries where the projects originate more comprehensively. They point out that most policy transfer research focuses "almost entirely on receivers of information"; criticising the lack of systematic comparative research on senders and producers. A second research gap can be found in the predominant focus of most studies on transfers between countries in the Global North. Policy transfer research on conditions related to the transfer of experiences between countries or cities of the Global North and South is rare (Campbell, 2012; Dolowitz and Medearis, 2009).

Through the analysis of the processes and conditions for transnational urban learning this study also aims to strengthen policy transfer research at the subnational level. Cooperation and exchange of experiences at the subnational level has been largely neglected by policy transfer research as Dolowitz and Marsh (2000, 12) stress: "Furthermore, while it is seldom examined, it should be stressed that when drawing lessons from other nations, actors are not limited to looking at national governments, but can, and do, look to other sub-national levels and units of government". Campbell (2012) and Medearis and Dolowitz (2013) confirm the gap in research on the processes and conditions related to urban learning and policy transfer.

The lack of research on policy transfer theory related to subnational North-South cooperation suggests an explorative and inductive approach towards theory development. In order to determine the specific conditions of knowledge exchange in transnational urban cooperation, the study locates urban partnerships within transnational climate governance theory and draws additional hypotheses from theoretical literature on the role of institutions, policy entrepreneurs and social capital.

3.2 Transnational Climate Governance Networks

Andonova, Betsill and Bulkeley's (2009) typology of transnational climate change governance networks offers a good starting point for classifying transnational urban climate cooperation (see table 1). In their typology, Andonova et al. address the gap in research on transnational governance frameworks and aim to start mapping the "patchwork of transnational governance networks" that has emerged beyond the formal international climate negotiations (ibid., 59)[34]. Andonova et al. refer to Risse-Kappen's (1995: 3, as cited by Andonova et al., 2009, 54) definition of transnational relations as "regular interactions across national boundaries when at least one actor is a non-state agent or does not operate on behalf of a national government or an international organization." The work points to the fact that Risse-Kappen's definition of transnational relations also includes cross-country interactions by governmental actors, "when at least one actor pursues her own agenda independent of national decisions." (Risse-Kappen, 1995, 9, as cited by Andonova et al., 2009, 54) Based on these qualifications Andonova et al. (2009, 56) define transnational governance networks as "cross-border networks of different configurations of actors" that "authoritatively steer constituents towards public goals".

---

34 A similar typology has also been developed by Bäckstrand (2008).

3.2 Transnational Climate Governance Networks

Type of Actors / Function	Public	Hybrid	Private
Information Sharing	UK-California initiatives	The Climate Group (TCG)	Pew Business Environmental Leadership Council
Capacity building and implementation	Cities for Climate Protection (CCP)	Renewable Energy and Energy Efficiency Partnership (REEEP)	World Business Council for Sustainable Development (WBCSD)
Rule setting	Regional Greenhouse Gas Initiative (RGGI)	Chicago Climate Exchange (CCX)	The Gold Standart

Table 1: Typology of Transnational Climate Change Governance Networks; source: Andonova, Betsill and Bulkeley, 2009, 60.

The typology of transnational climate change governance networks classifies networks, according to their dominant actor structure and institutional set-up, as either public, private or hybrid public and private transnational networks; as well as according to the governance functions that such networks embody, such as information sharing, capacity building/implementation, and rule-setting. The typology lists one practical example for each combination of institutional forms and functions (see table 1).

Andonova et al. do not claim that the typology is exhaustive or fully covers the multiplicity of transnational governance networks that exist in the field of climate change, urging future research to add more examples.

Transnational urban climate partnerships, the subject of analysis in this study, fulfill the requirements of transnational governance networks as defined above. Andonova et al. explicitly mention subnational authorities, such as cities, regions and states as actors participating in transnational climate governance networks. In fact, all three examples which the typology lists under "public" transnational collaboration include subnational governments, namely the climate cooperation between the U.S. state of California and the United Kingdom (UK-California Initiative), the CCP network of 600 local governments worldwide initiated by ICLEI, and the Regional Greenhouse Gas Initiative (RGGI) in which ten U.S. states and two Canadian Provinces set up a CO_2-emissions cap and trade program. In these examples, subnational governments have voluntarily agreed to pursue climate protection activities and to exchange experiences with their network partners. The initiatives exemplify the emergence of new climate governance arenas beyond the formal negotiations of the UNFCCC framework and highlight the role that subnational governments can play in fostering global climate protection efforts outside of the "boundaries of formal intergovernmental diplomacy" (Andonova et al., 2009, 61).

This study analyses transnational urban partnerships which take differing institutional forms, such as predominantly public, predominantly private and hybrid public and private partnerships, as outlined in Andonova et al.'s typology. The typology does, however, not make any assumptions about the effectiveness and impact of different institutional forms of transnational governance networks. This study therefore refers to the theory of institutionalism which offers more guidance in this regard, as demonstrated in Ralston's (2013) study on subnational cooperation between German and U.S. states.

3.3 New Institutionalism

Key questions regarding the institutional set-up of transnational urban partnerships refer to the benefits and shortcomings of different institutional forms of urban climate cooperation and whether any conclusions can be drawn about the effectiveness of predominantly public or private partnerships or mixtures thereof.

Institutionalism literature generally takes structure as a starting point for the analysis of political processes. Peters (2012, 174) states that the key assumption binding all approaches of institutionalism is that structure determines behaviour and agency as well as the outcomes of political processes: "The most fundamental point is that scholars can achieve greater analytic leverage by beginning with institutions rather than with individuals."

Young (2002) specifies that institutionalism studies analysing global environmental change aim to deepen the understanding of the roles that institutional drivers play in fostering environmental change. He explains that most scholars in this domain take the perspective of the "new institutionalism" movement in the social sciences and would agree to the definition of institutions as "sets of rules, decision-making procedures, and programs that define social practices, assign roles to the participants in these practices, and guide interactions among the occupants of individual roles." (ibid., 5). In contrast to more traditional institutionalist approaches that primarily focus on formal institutional design, new institutionalism research analyses rules and social practices that are actually in use. Beyond this common understanding of institutions, new institutionalism literature is too diverse and fragmented to offer general propositions about how to design effective institutions, Young points out (ibid.).

While stating that it is difficult to draw general conclusions, Young discusses the conditions for effective institutionalisation in the specific field of environmental governance at the subnational level, analysing local environmental regimes and their linkages and interaction with their institutional environment. Young argues that in complex societies the success of local environmental regimes depends on two factors. Firstly on their local institutionalisation which needs to be balanced between strength and depth on the one hand to influence behavioural change; and flexibility on the other hand to reach high compliance by its subjects. Moreover a local environmental regime should be designed in a way which reduces conflicts with existing local institutions, provides financial incentives and ensures it is considered legitimate by those affected by it. Secondly, Young emphasises that the success of local environmental regimes is conditional upon their vertical integration into the institutional environment of higher policy levels such as subnational and central state institutions. Young refers to the concept of comanagement which highlights the relevance of working partnerships between local and (sub)national institutions, as (sub)national authorities have important formal decision-making competencies and resources for the implementation of local projects (ibid.).

According to Young, the relevance of the institutional interplay for the design and performance of local environmental regimes is often neglected by the research community as it is very difficult to research. However, he argues that as institutional density increases, institutional interplay is gaining importance and has increasing influence on institutions' performance as well as its robustness (ibid.).

Ralston (2013) demonstrates in one of the first in-depth comparative studies on this topic that the propositions from new institutionalism can also be applied to transnational cooperation at the subnational level. Ralston analyses sustainability partnerships between German and U.S. subnational states, comparing enabling factors and barriers to successful cooperation. Referring to New Institutionalism and multi-level governance[35] literature, Ralston concludes that transnational sub-

---

35 Ralston draws on the two types of multi-level governance (MLG type I and MLG type II) developed by Hooghe and Marks (2003) with MLG type I representing more formalised governance forms that are usually characterised by non-overlapping state jurisdictions (e.g. federalist structures) and MLG type II embodying new governance forms that are more flexible and less formalised; are developed by interactions between both state and non-state actors; and often span several jurisdictions. According to Hooghe and Marks the two types of MLG co-exist, with MLG type II often being embedded into the legal structures of MLG type I. Ralston identifies transnational subnational cooperation as a typical form of MLG type II and concludes that with regard to subnational partnerships MLG type II has in fact to be embedded into the legal structures of MLG I to function effectively.

national partnerships need to be formally institutionalised into the state legal system in order to ensure long-term sustainability and reduce their dependency on partnership champions. Ralstons findings are also supported by Rose's (1993) study of policy transfers which argues that a program can only be transferred across countries if it is stated in law.

This study discusses whether the benefit of state institutionalisation is also evident in transnational urban partnerships. It investigates what forms of institutionalisation lead to the success or failure of a collaboration's outcome, and whether transnational urban partnerships benefit from close interlinkages with local, subnational and central state institutions. The following hypothesis is tested in the comparative case-study analysis:

Hypothesis 2: The more a transnational urban partnership project is institutionalised into the state system, the more likely the project is to succeed.

Hypothesis 2 is also tested with the help of an index system (see chapter 4). This methodological approach is recommended by Young (2002, 14) who states that the "question of [institutions'] performance requires the specification of criteria of evaluation, followed by an assessment of the extent to which actual outcomes measure up in terms of those criteria."

The two remaining hypotheses derived from theoretical literature highlight the role of human agency and relations in governance processes, an area which tends to be neglected in institutionalism literature. Young states that due to its primary focus on the influence of institutional drivers, institutionalism provides a crosscutting theory for analysing global environmental change. However Young also points out that while institutions play a significant role in most environmental changes involving human action, they are seldom the only cause. Young (ibid., 4) states that a crucial task for institutionalism research remains "to separate the signals associated with institutional drivers from those associated with other drivers, and to understand how different driving forces interact with each another to account for observed outcomes."

The focus on additional driving forces, and in particular the role of individuals and networks in environmental governance, is crucial for the analysis of Indian-German urban partnerships. As Wagner (2006) concludes in his analysis of the political system in India, governing in India in fact often centres more around individuals than institutions. This study therefore borrows two theoretical concepts to help understand the role and impact of agency and human relations in transnational urban partnerships.

3.4 Policy Entrepreneur Concept

This work refers to the policy entrepreneur concept as a framework for the study of agency in transnational urban partnerships. The concept of policy entrepreneurs has been promoted by John W. Kingdon (2011), even though, as he himself points out, the term and concept were developed in earlier research such as Salisbury (1969) and Walker (1974, 1981). Kingdon (ibid., 122) defines policy entrepreneurs as "advocates for proposals or for the prominence of an idea" within a policy community. A similar definition is provided by Michael Mintrom (1997, 739) who describes policy entrepreneurs as "political actors who promote policy ideas" and "people who seek to initiate dynamic policy change".

Kingdon (2011, 122) explains that policy entrepreneurs can be state or nonstate actors that have gained formal or informal authority through elected or appointed positions or through their personal charisma. A major common characteristic of policy entrepreneurs is "their willingness to invest their resources – time, energy, reputation, and sometimes money – in the hope of a future return." Kingdon states that policy entrepreneurs can be driven by diverse motivations, including concrete targets such as fostering the introduction of a desired policy; by personal values such as civil and political engagement; or private interests such as job security or career development.

According to Kingdon, policy entrepreneurs are crucial actors in the development and enforcement of public policies or projects. He points out that they often play important roles in identifying and utilising policy windows: "Policy entrepreneurs play a major part in the coupling at an open policy window, attaching solutions to problems, overcoming the constraints by redrafting proposals, and taking advantage of politically propitious events." (ibid., 165-166) Kingdon compares policy entrepreneurs to persistent and skilled navigators, who are "ready to paddle, and their readiness combined with their sense for riding the wave and using the forces beyond their control contributes to success." (ibid., 181-182)

Successful policy entrepreneurs often embody the certain characteristics. Firstly, they show extraordinary levels of engagement and persistence in the promotion of their targets and invest large quantities of their personal resources (Kingdon, 2011).

Secondly, policy entrepreneurs need negotiating skills and the ability to convince others of their ideas. Successful policy entrepreneurs are able to reframe policy issues and present and sell their arguments in different ways to different audiences, which is "clearly a task for the politically savvy" (Mintrom, 1997, 740). In order to form effective coalitions policy entrepreneurs need team building skills as well as the will and ability to conduct leadership in these teams. Mintrom points out that those policy entrepreneurs that have past experience of successful coalition-building are more likely to be able to build effective coalitions again (ibid.).

Thirdly, in order to build such coalitions, policy entrepreneurs require good access to diverse policy networks. Mintrom (1997) and Mintrom and Vergari (1998) argue that it is crucial for successful policy entrepreneurs to be well connected to key members of the policy community within their own jurisdiction (internal networks) as well as with actors beyond their jurisdictions (external networks). Policy entrepreneurs require access to internal networks in order to discover the diverse interests and ideologies of the policy-making community and to establish contacts that improve the policy entrepreneurs' credibility. Their credibility is further strengthened by having good access to external policy and expert networks, which are also an important source for policy entrepreneurs to explore, share and discuss innovative ideas (ibid.). Mintrom and Vegari (1998) find that in agenda-setting policy entrepreneurs tend to make use of both external and internal networks, while when gaining approval for a desired project or policy they depend mainly on their internal network.

Assessing the overall impact of policy entrepreneurs, Mintrom (1997) concludes that their involvement can significantly increase the likelihood of introducing and implementing policy innovations. Mintrom (ibid., 740) draws a parallel with market entrepreneurs, suggesting that policy entrepreneurs are "able to spot problems, they are prepared to take risks to promote innovative approaches to problem solving, and they have the ability to organise others to help turn policy ideas into government policies."

This study investigates the role policy entrepreneurs play in the successful implementation of transnational urban partnerships. It explores to what degree such partnerships are driven by or indeed depend on engaged local individuals from the partner cities. This study proposes and adopts the term 'partnership entrepreneurs' to describe policy entrepreneurs that engage in urban cooperation (or any other form of transnational or international cooperation). Based on the propositions about the characteristics, abilities and impact of policy entrepreneurs from the theoretical literature outlined above, the following hypothesis is derived and tested in the analysis of German-Indian urban climate partnerships:

Hypothesis 3: A transnational urban partnership project is more likely to succeed if it is driven by engaged, persuasive and well-networked partnership entrepreneurs.

The partnership entrepreneur must have personal conviction in the partnership project and be willing to invest personal resources in the initiative. The individual must also be able to convince other stakeholders and form effective coalitions, therefore allowing them to access policy networks within and beyond their home or partner city (for more details of how indicators were developed see section 4.3).

To further explore the role of human relations and networks in bridging the distinct socio-cultural, political and economic contexts of Indian-German urban climate partnerships, this work also draws on the concept of social capital.

3.5 Social Capital Theory

The concept of social capital in the empirical political sciences was proposed by Robert D. Putnam in the early 1990s (Heydenreich-Burck, 2010). Together with his co-authors Putnam conducted several renowned studies that analyse the impact of social capital in different socio-economic contexts (Putnam 2000, 2002; Putnam, Feldstein and Cohen, 2003; Putnam, Leonardi and Nanetti, 1994). Referring to earlier applications of the term by Loury (1977, 1987), Coleman (1990) and Ostrom (1992), Putnam et al. (1994, 167) define social capital as "features of social organization, such as trust, norms, and networks, that can improve the efficiency of society by facilitating coordinated actions".[36]

The concept's main assumption is that social capital is highly beneficial to all kinds of joint projects at the local level as it serves as a response to the dilemmas of collective action, so-called "social dilemmas", such as the tragedy of the commons, the free rider problem or the prisoner's dilemma.[37] Putnam et al. state that according to game theory these dilemmas would foster uncooperative behavior, arguing that in reality, however, uncooperative behavior is less common than game theory's concept of rationality would predict. In fact, they find many examples of cooperative societal solutions to collective action challenges and they argue that social capital is the foundation of these solutions. Putnam et al. further highlight that thanks to spill-over effects the whole community benefits from social capital development in individual projects (ibid.).

Referring to Putnam's conceptual framework, Heydenreich-Burck (2010) adds that social capital also improves the performance of governance processes as it reduces transaction costs as it leads to a) higher compliance with norms, thereby

---

36 Putnam and Goss (2002) explain that the term 'social capital' was in fact already introduced in a study by Lyda Judson Hanifan (1916, 130) who defined social capital as "(...) good will, fellowship, sympathy, and social intercourse among individuals and families who make up a social unit." (as cited by Putnam & Goss, 2002, 4). Heydenreich-Burck (2010) explains that the basic idea of the concept is even older referring to Tocqueville (1987) who in 1840 published a study about the social-integrative role of voluntary organisations.

37 For an overview of the social dilemmas see Dawes (1980).

reducing the resources required for legislation and surveillance; b) smoother decision-making through the reduction of partisan blockades; and c) less polarised preferences of citizens.

According to Putnam et al. (2003, 9) social capital is "necessarily a local phenomenon", as "trust relationships and resilient communities generally form through local personal contact". While there is a multiplicity of studies on social capital in diverse local contexts, the concept has not been comprehensively researched with regard to urban learning and cooperation.

One exception is Campbell (2012) who elaborates on the significance of social capital among citizens and stakeholders in the facilitating of internal and external exchange in cities. Campbell argues that in globalizing cities, urban learning is increasingly driven by non-state actors through "informal leadership networks" (ibid., 11). Campbell explains that these informal networks are based on trustful relations between business, civil society and youth leaders ("clouds of trust"). He concludes that such relationships are the "gold standard of learning cities", as they have a highly important stabilising function in urban exchange and learning processes by guaranteeing continuity in times of political leadership change (ibid., 14-15).

Without explicitly referring to social capital, several studies pointed to the relevance of analysing the role of personal and equal relationships, trust and inclusive approaches as prerequisites for the establishment of transnational urban cooperation. Bontenbal and van Lindert (2008, 477) state that "feelings of mutual friendship and trust" are crucial conditions for the setting up and maintaining of city partnerships. Van Ewijk and Baud (2009) highlight the need for equality, mutual trust, the appreciation of diversity and openness as important elements of urban North-South cooperation.

In the case of urban South-North collaboration, as for example in Indian-German city partnerships, the protagonists face the challenge of building collaborations despite the significant contextual and cultural differences that shape their cities. It can be assumed that in particular the development of what Putnam et al (2003, 2) define as "bridging social capital" is a crucial prerequisite for the creation and implementation of transnational urban partnership projects. Putnam et al. distinguish between "bridging" and "bonding" social capital, stating that "bridging" social capital is created within networks of people with diverse socio-economic backgrounds which tend to be more outward-looking. Functioning as a "sociological WE-40" or lubricant, the existence of such heterogenous networks are crucial to developing resilient and inclusive communities, Putnam et al. argue.[38]

---

38   "Bonding social capital" on the other hand evolves in networks of people who have similar socio-economic backgrounds and tend to be more inward-looking. "Bonding social capital" thus

They point out that although "bridging" social capital is often more difficult to achieve than "bonding" social capital, it is vital for facilitating cooperation in increasingly complex societal settings (ibid.).

This study investigates the role social capital plays in Indian-German city partnership projects. Based on the assumptions of social capital theory and studies on city cooperation on the topic, the following hypothesis is tested in the comparative case study analysis:

Hypothesis 4: The more social capital protagonists develop as part of the transnational urban partnership, the more likely the partnership project is to succeed.

To design the index measuring social capital in transnational urban partnerships the study refers to dimensions of social capital as developed by World Bank researchers, complementing these with indicators on intercultural competence and equality that are of particular relevance for urban North-South cooperation (see index system in section 4.3).

---

strengthens the social cohesion of a network, as a kind of "sociological Super-Glue" (Putnam et al., 2003, 2-3).

4 Methodology

4.1 Addressing the Northern Bias in Urban Research

This study aims to contribute to the development of more inclusive international urban research. It supports the demands of post-colonial researchers for a greater inclusion of Global South experiences in theory development in general (Said, 1994; Connell, 2007) and more specifically in urban research (McFarlane, 2010; Robinson, 2002, 2006, 2011, Ward 2008).

McFarlane (2010) criticises the lack of internationally comparative urban research, the majority of which is limited to comparative studies within and between Europe and North America and the analysis of "global cities", i.e. large and influencial cities, such as London, New York and Tokyo. He points out that cities from the Global South are generally underrepresented in urbanism, as are less prominent "second-tier" cities. McFarlane identifies an eminent research gap in the form of comparisons between cities from different world regions and at different stages of economic development.

4.1.1 Assumed Incommensurability

As a prerequisite for more "cosmopolitan" urbanism, researchers in the field of post-colonial studies urge that urban studies must overcome the notion of incommensurability of cities in the Global North and South. Robinson (2011) elaborates that most urbanism scholars acknowledge the problem that there is a bias in existing urban theory towards experiences from a limited number of rich Northern cities. When conducting research, however, strict assumptions about the incommensurability of cities with different socio-economic contexts, political systems and cultural traditions remain dominant.

Post-colonial urbanism scholars counter this view arguing that it is precisely these rigorous perspectives on (in)commensurability that have contributed to the systematic neglect of Southern cities' experiences in urban studies. They criticise the lack of conceptual discussions within urban research on how to best compare cities with differing contextual conditions, beyond the standard categories of North/South, developed/underdeveloped, rich/poor, (post)-socialist/capitalist, etc.

(McFarlane 2010, Robinson 2011). Robinson (2011) highlights that urbanism research needs to tackle the challenge of developing a more internationally inclusive orientation; not only to address constrained and biased definitions and concepts which contradict urbanism's claim to universality, but also to acknowledge the reality of globalisation and its resulting economic, political and cultural connections and dependencies between cities worldwide. Robinson identifies a contradiction between current public policy discourse that (possibly too promptly as Robinson points out) promotes best practise benchmarking and learning, and the hesitant approach of the research community which has so far widely refused to undertake globally inclusive city comparisons.

4.1.2 Northern Bias Reflected in Urban Climate Change Research

The debate about refocusing urbanism towards more comparative research involving both cities from the Global North and the Global South is exemplified in research on cities and climate governance (see section 2.1). Several leading cities and climate change scholars, such as Harriet Bulkeley, David Satterthaite, Ingemar Elander and Susan Parnell confirm the need to address the gap in international comparative urban climate research and highlight in particular the need to strengthen the representation of Southern cities' experiences.39

4.1.3 Focus on Urban Processes and City Connections as a Way Forward

According to Robinson (2011, 14) a promising strategy to overcome the assumed incommensurability of cities could be to look at urban processes rather than the city as a territorial entity. She explains that

> there are many urban processes for which neither formal administrative boundaries nor the functional regions of cities would be the relevant scale for comparison. Instead, processes that exceed a city's physical extent — circulations and flows — as well as phenomena that exist and operate at a smaller scale than the city should be the relevant units for comparison.

Robinson specifies that in particular the analysis of connections between cities could serve as a starting point for more cosmopolitan urban research, as interrelations between cities could be investigated and compared without the cities having to be categorised:

---

39 In discussions as part of the workshop "Mediating Climate Change in the City. Experimenting with Urban Responses" at Durham University, March 19-23, 2012.

Many phenomena in cities are tied into connections and flows that stretch beyond the city's physical or territorial extent and that entrain other urban contexts into the dynamics of that city. In a further step, these connections themselves become the units of comparisons. (...) More generally, as cities are already interconnected, different ends of the connections might be brought into stronger analytical and not simply empirical relation. (...) Not only do the two ends of a connection come into view, though; the connections themselves might well form the focus of comparison. The connection, then, becomes the case. (ibid.)

This study supports the arguments of proponents of a more cosmopolitan form of urbanism. Via a comparative case study analysis of the connections between cities from different contexts (India and Germany) this research project aims to contribute to the strengthening of internationally inclusive approaches to methodology and theory development in urbanism and urban climate governance research.

4.2 Comparative Case Study Analysis

Robinson (2011) argues that qualitative case studies are particularly well suited to account for the diversity of cities and to gain a comprehensive situative understanding of their distinct contexts, actor constellations and political processes. This study supports Robinson's conclusion reflected in the choice of a qualitative research design for the work.

4.2.1 Qualitative Case Study Analysis

Robert K. Yin (2009, 2) explains that in comparison to quantitative approaches, qualitative case studies are better able to address multifaceted social processes, particularly when "(a) 'how' or 'why' questions are being posed, (b) the investigator has little control over events, and (c) the focus is on a contemporary phenomenon within real-live context." George and Bennett (2005, 19) specify that qualitative case studies are a proven method for the identification of causal mechanisms and causal interlinkages in complex settings; making them therefore a crucial means of generating new and testing existing theory:

> Case studies are generally strong precisely where statistical methods and formal methods are weak. We identify four strong advantages of case methods that make them valuable in testing hypotheses and particularly useful for theory development: their potential for achieving high conceptual validity; their strong procedures for fostering new hypotheses; their value as a useful means to closely examine the hypothesized role of causal mechanisms in the context of individual cases; and their capacity for addressing causal complexity.

The ability of qualitative case studies to investigate causality in complex social processes is crucial in the analysis of the factors explaining success and failure in Indian-German urban climate partnerships. This study is therefore based on a qualitative research design. Quantitative data collection methods are not applied, as they are less capable of identifying the multiplicity of variables that determine processes of transnational urban climate cooperation. Moreover, the amount of existing German-Indian climate partnerships and the number of people engaging in these partnerships are too limited to justify the application of statistical methods.

Consequently, this study cannot provide statistically proven answers and solutions. Yin emphasises that this can never be the task of qualitative case study analyses. He points out that, rather than aiming for "statistical generalization", the purpose of conducting detailed case studies is to conduct "analytical generalization", by testing, modifying and generating theory (Yin, 2009, 38).

4.2.2 Multiple Case Study Analysis

Both George and Bennett (2005) and Yin (2009) state that in principal it is possible to develop theory on the basis of single case studies (thereby contradicting the arguments of for example King, Keohane and Verba, 1994). George and Bennett (2005) and Yin (2009), however, emphasise the benefits of setting up multiple case study designs. An analysis of more than one case would facilitate the exploration of patterns across the cases and thereby enable a direct testing of the generated theory as part of the same study. This would strengthen the analytical generalisability of the findings. Yin (2009) therefore recommends that, whenever possible, at least two cases should be included, with each additional case further improving the external validity of the findings.

In the study of Indian-German urban partnerships this argument is even more relevant, as there is no basis of extensive preliminary research with which new findings can be compared. By conducting four case studies, the diversity of city collaboration approaches can be better represented than with a single case study approach.

Moreover, a multiple case study design reduces the risk of not being able to collect the necessary data required for the study. Or, as Yin (2009, 61) points out: "Single-case study designs are vulnerable if only because you will have put 'all your eggs in one basket'." A multiple case study approach provides the researcher with a certain degree of flexibility to adjust the research design and still allows for the realisation of the study even if the researcher cannot collect sufficient data for one of the selected cases and is forced to exclude it from the study. The multiple case study approach was shown in the conducting of this study to be beneficial for exactly this reason. The study of Indian-German partnerships required initial ground research in both countries to identify such partnerships and verify the actual extent of the projects' implementation that are mentioned in existing literature or databases. In fact, one previously selected case, the city partnership between Coimbatore (India) and Esslingen (Germany), was excluded from the study after the initial research stay, as the setting up of the collaboration was behind schedule and no concrete projects had been planned. This case was then replaced with the collaboration between Nashik and Hamburg, which was only identified in explorative expert interviews conducted during the author's first research stay in India.

The comparison of multiple cases is also supported by post-colonial urbanism researchers, such as Robinson (2011, 6), who states that multiple case studies increase the potential to challenge existing theory and identify new dynamics and research questions:

> In relation to urban studies, it has been particularly productive to bring the experiences of different case-study cities into careful conversation with one another in order to reflect critically on extant theory, to raise questions about one city through attending to related dynamics in other contexts, or to point to limitations or omissions in existing accounts.

4.3 Research Approach

The question of how to address and incorporate existing theory in a research design is contested among the proponents of qualitative case study approaches.

Several authors of case study text books, such as for example van Evera (1997), George and Bennett (2005) and Yin (2009), generally demand a more rigorous founding of case studies in existing theoretical literature in order to relate them to on-going academic debates and to strengthen the validity and comparability of their findings.

In contrast, many post-colonial scholars highlight the need for keeping more critical distance from existing theory. They call for new approaches to global theory development in order to overcome the Northern bias in the literature (Said 1994, Chakrabati 2002, Connell 2007) and especially in urbanism (McFarlane, 2010; Robinson, 2002, 2006, 2011; Ward, 2008). Opinions on how to achieve this reorientation range from radical demands for an entire detachment (Robinson, 2002) and suggestions for modifications to existing theory (Denters and Mossberger 2006, McFarlane 2010), to arguments for generally more plurality in theory development (Nijman 2007, Ward 2008).

Acknowledging the benefits of both arguments, this study suggests finding a middle way by critically reviewing, adapting and complementing existing theoretical concepts in research designs that include experiences of both Southern and Northern regions. This study takes therefore a combined inductive and deductive approach. Based on the experiences of initial research visits and interviews in German and Indian cities, four research hypotheses have been drawn from several complementary theoretical concepts (see chapter 3). The hypotheses have served as guidelines for the collection of data for the case studies and will be revised based on a comparative analysis of the results. To operationalise the hypotheses an index system has been established, in which propositions from the existing literature are complemented by empirical findings.

4.4 Development of Index System to Operationalise Dependent and Independent Variables

4.4.1 Approach

The index system has been developed with reference to recent efforts by the research community to minimise the gap between measurement tools and the complex reality of the social world by incorporating and integrating theoretical and empirical insights (Gale, Heath, Cameron, Rashid, and Redwood (2013); Tjandradewi and Marcotullio (2009); Tucker (2010)).

Tucker (2010) explains that the opportunity to include and adjust indicators based on initial empirical research increases the applicability of index systems across different global regions and times, thereby providing a strategy for decentralising research concepts from the Global North. Adopting this approach makes it even more important to clearly define the scope of a "grounded" index system in a concrete study, as this in turn defines the scope of analytical generalisation. This study's index system has been built to explore transnational South-North urban cooperation in the area of climate protection and low carbon development (as defined above, see section 1.1.2). As it was designed on the basis of experiences gained from initial field visits and interviews it partly reflects the contextual conditions of Indian-German city partnerships. In particular the applicability of the indicators that were included on the basis of empirical analysis in Indian and German cities should be carefully reviewed before applying the index system to an analysis of urban cooperation involving cities from other countries.

The index system includes one index measuring the dependent variable of partnership project success, plus four indexes for the independent variables which are the conditions for success and failure in the projects: 1) the knowledge exchange strategy, 2) institutionalisation into the state system, 3) partnership entrepreneur, and 4) partnership social capital (see research hypotheses in section 1.3).

Due to the lack of references in existing literature on the weighting of indicators, the study opts to score them according to the same logic designed by the author (positive score=2 points, medium/neutral score=1 point, negative score=0 points). Where applicable and reasonable, indicators are scored individually for both partnership cities (these indicators are marked with "→ *individual scoring for both cities*", see below). This allows for a more sophisticated assessment of cities' distinct contexts and approaches to city partnerships. The results are then aggregated, for the total indicator scores as well as for the overall index scores.

In the case studies, the hypotheses are tested on the basis of the scores of the independent variable indexes and their correlation with the scores of the project's success index. Positive correlations (high scores in both the independent variable index and the success index, or low scores in both the independent variable index and the success index) strengthen the hypotheses, negative correlations (high scores in the independent variable index and low scores in the success index, or vice versa) weaken their validity.

To strenthen the reliability of the index system and the scoring, a sample of the data (the DEWATS case study) has been test scored by a second evaluator. As the scores are largely congruent, no modifications have been conducted in the index system.

4.4.2 Index System

Success Index[40] (7 indicators)
(1) Implementation achievements [no implementation = 0, micro-level implementation (small-scale/pilot project implementation) = 1, macro-level implementation (large-scale/city-wide project implementation) = 2]
(2) Budget and schedule performance [budget *and* schedule exceeded = 0, budget *or* schedule exceeded = 1, budget and schedule compliance = 2]

---

40 The success index builds on previous studies' suggestions for the development of success indicators for sub-national transnational cooperation (Bontenbal, 2009) and (Ralston, 2013). Indicators (1), (2), (4), (5) and (7) are oriented on the evaluation criteria developed by the OECD DAC Network on Development Evaluation (2010, 13-14) which were created in order to evaluate development cooperation projects. To allow for a more sophisticated evaluation of transnational urban partnership programs two additional indicators accounting for capacity building (3) and partnership mutuality (6) have been added. Indicator (3) was derived from Bontenbal (2009, 182), Johnson & Wilson (2009, 216) and Devers-Kanoglu (2009, 202-204) and indicator (6) from Johnson & Wilson (2009), Bontenbald & van Lindert (2008, 479), Bontenbal (2009, 188) and Ewijk & Baud (2009, 218).

(3) Local capacity building [no capacity building (individual learning, improvement of administrative processes, institutional reform, establishment of local knowledge centre) = 0, some capacity building (in the forms listed above) = 1, extensive capacity building (in the forms listed above) = 2] → *individual scoring for both cities*
(4) Benefits for target groups [no benefits for target groups = 0, benefits for some target groups = 1, benefits for most/all target groups = 2]
(5) Impact beyond the partnership project [no impact = 0, some impact = 1, broad impact = 2]
(6) (Perceived) partnership mutuality [perceived one-sided learning and benefits = 0, some perceived mutuality = 1, perceived mutuality = 2]
(7) Post-project sustainability of the city partnership [no post-project activities = 0, some post-project activities = 1, extensive post-project activities = 2]

Index 1: Knowledge Exchange Strategy[41] (7 indicators)

    a. *Prior program and context evaluation*
       (1) Prior evaluation of how programme operates in practice in original setting [no evaluation = 0, limited evaluation = 1, comprehensive evaluation = 2]
       (2) Prior evaluation of contextual conditions for partnership project [no evaluation = 0, limited evaluation = 1, comprehensive evaluation = 2]

    b. *Knowledge exchange process*
       (1) Partnership project followed a systematic approach to the cooperation and exchange of knowledge [no systematic approach = 0, systematic approach at times = 1, systematic approach throughout whole project = 2]
       (2) Continuity of interaction between partnership actors throughout knowledge exchange process [occasional interaction = 0, regular interaction in some phases of programme transfer = 1, regular interaction throughout entire programme transfer = 2]

---

41   Indicators a(1)-(2) are derived from Dolowitz & Marsh (2000, 17), b(1) from Mossberger & Wolman (2003, 438), Tews (2008, 80) and Campbell (2012, 187), b(2) from Medearis & Dolowitz (2013) and Ansell & Gash (2007, 543), b(3) from Stone (2000) and Mossberger & Wolman (2003, 438), c(1) from Kingdon (2011, 166ff), Tews (2008, 80) and Dolowitz & Medearis (2009, 694) and c(2) Betsill & Bulkeley (2007, 452).

4.4 Development of Index System to Operationalise Dependent and Independent Variables

(3) Involvement of external partnership moderator/facilitator [no involvement = 0, occasional involvement = 1, extensive involvement = 2]

c. *Political strategy*
(1) Utilisation of policy windows for partnership project [no policy window = 0, partial policy window = 1, policy window = 2]
(2) Partnership project addresses intrinsic interests of partnership actors [no intrinsic interests = 0, some intrinsic interests = 1, strong intrinsic interests = 2] → *individual scoring for both cities*

Index 2: Linkages with State Institutions[42] (6 indicators) → *all Index 2 indicators: individual scoring for both cities*

a. *Formal approval of partnership project by state institutions*
(1) Formal approval of partnership project by state institutions (local) [no state approval = 0, partly state approved = 1, fully state approved = 2]
(2) Formal approval of partnership project by state institutions ((sub)national) [no state approval = 0, partly state approved = 1, fully state approved = 2]

b. *Provision of state resources for the partnership project*
(1) Partnership project is executed by state employees [no state employees = 0, partly state employees = 1, primarily state employees = 2]
(2) State funding for partnership project [no state funding = 0, part state funding = 1, primarily state funding = 2]

c. *Leadership of state actors in the partnership project*
(1) State institution as project coordinator [non-state coordination = 0, part state coordination = 1, full state coordination = 2]

---

42 Indicators a(1)-(2) are derived from Young (2002, 28-32, 111), Berkes (2002, 293-294) and Ralston (2013), a(2) from Nakamura (2010, 2) and Tjandradewi & Marcotullio (2009, 168), b(1) from Schwedler (2011, 49), b(2) from Ralston (2013, 296-298) and Schwedler (2011, 49), c(1) induced from author's field observations during research stays in India, c(2) from Nakamura (2010: 15), Campbell (2012, 189, 197) and Tjandradewi & Marcotullio (2009, 168).

(2) Commitment of local political leaders (commissioners/mayors) to the partnership project [no commitment = 0, some commitment = 1, full commitment = 2]

Index 3: Local Partnership Entrepreneurs[43] (6 indicators) → *all Index 3 indicators: individual scoring for both cities*

a. *Personal Engagement and Persuasiveness*
  (1) The partnership entrepreneur (PE) is convinced of the feasibility and benefits of the partnership initiative [not convinced = 0, partly convinced = 1, strongly convinced = 2]
  (2) The PE invests his/her own resources (time, reputation, money) in the partnership initiative [no investment of private resources = 0, some investment of private resources = 1, extensive investment of private resources = 2]
  (3) The PE is able to convince different stakeholders of the viability of the partnership project [lack of ability to convince = 0, some ability to convince = 1, strong ability to convince = 2]

b. *Personal Contacts and Access to Policy Networks*
  (1) The PE has established good relations with members of the internal policy network of his/her city [lack of internal network = 0, small internal network = 1, large internal network = 2]
  (2) The PE is well networked in the external policy community beyond his/her own city [lack of external network = 0, small external network = 1, large external network = 2]
  (3) The PE is well connected to actors from the partner city [no connections with partner city actors = 0, some connections with partner city actors = 1, extensive connections with partner city actors = 2]

---

43 Indicator a(1) is derived from Ralston (2013, 14-15), a(2) from Kingdon (2011, 122-123, 180-181), a(3) from Mintrom (1997, 740, 766-767), b(1) and b(2) from Mintrom (1997, 766-767) and b(3) from author's own research (interview with AFG, 26-11-2012)).

4.4 Development of Index System to Operationalise Dependent and Independent Variables 71

Index 4: Partnership Social Capital[44] (7 indicators)

a. *Collective Action, Trust and Equality*
   (1) Experiences from prior collective action [no prior partnership project = 0, one prior partnership project = 1, more than one prior partnership project = 2]
   (2) Trust in the partner city actors' reliability and competences [lack of trust = 0, limited trust = 1, trust = 2] → individual scoring for actors from both cities
   (3) (Perceived) equality by German and Indian partnership actors [(perceived) inequality = 0, some (perceived) inequality = 1, (perceived) equality = 2] → *individual scoring for actors from both cities*

b. *Intercultural Competence and Communication*
   (1) Intercultural experiences: partnership actors' exposure to international exchange [no experience = 0, some experience = 1, extensive experience = 2] → *individual scoring for actors from both cities*
   (2) Intercultural communication [major communication barriers = 0, some communication barriers = 1, no communication barriers = 2]

---

44   The Social Capital Index is based in part on the six dimensions of social capital developed in a work by a group of researchers under the head of the World Bank between 2000 and 2006 (Nyhan Jones & Woolcock 2010, 546-555). The 6th dimension they name, "Empowerment and political action", is excluded from the Social Capital Index, as it is part of the dependent variable in this study and represented in the success index (indicators (3) and (4)). (The exclusion is indirectly proposed by Nyhan Jones and Woolcock themselves who emphasise that "Empowerment is a broader concept than social capital" (ibid, 554)). Each of the remaining five dimensions is represented by the indicators a(1)-(2), b(2) and c(1)-(2). They are complemented by two additional indicators measuring 'equality', and 'intercultural competence', derived from the existing literature and the author's own research. In addition to the World Bank index, references have been taken from the following authors: indicator a(1) is derived from Krishna & Shrader (2000, 20-25); a(2) from Putnam, Leonardi & Nanetti (1994, 170-171); a(3) from Johnson & Wilson (2009, 216), van Ewijk & Baud (2009, 220) and Mossberger & Wolman (2003, 432-433); b(1) from the author's own research (interviews with Forum Städtesolidarität Bremen-Pune and LAFEZ, 27-02-2013; (27-02-2013), Hamburg Wasser, 01-03-2013, and GIZ, 05-12-2012); b(2) from the author's own research (interview with AFG, 26-11-2012); c(1) from Campbell (2012, 11, 206) and Bontenbal (2009, 182, 186); and c(2) from Putnam & Feldstein (2003, 2-10).

c. *Stakeholder Involvement and Inclusion*
   (1) Involvement of local informal leadership networks in the partnership project [no involvement = 0, partial involvement = 1, intensive involvement = 2] → *individual scoring for both cities*
   (2) Citizens from diverse social backgrounds (wealth, social status, ethnicity, caste, religion, age) participate in the partnership project [similar backgrounds = 0, partly diverse backgrounds = 1, diverse backgrounds = 2] → *individual scoring for both cities*

4.5 Case Selection

Indian-German city partnerships are selected as case studies because urban centres in both countries are currently undergoing processes of transition which are of high relevance for addressing global climate change.

Many German cities are setting up comprehensive and large-scale climate and renewable energy strategies as part of Germany's energy transition. In fact, local and regional initiatives led by citizens and local governments have played a major role in facilitating the country's "Energiewende", the move towards a large-scale energy supply made up of renewable energy sources (Beermann and Tews, 2015). Countries and cities around the world are closely observing the implementation of Germany's ambitious target to transform its energy system, and research has started to explore the influence of the German energy transition as a model for local and national low carbon initiatives in other countries (Marquardt et al., 2015).

Indian cities are facing perhaps an even more fundamental transformation. Like many other cities from developing and emerging economies Indian urban centres are faced by the challenge by increasing urbanisation and industrialisation. Although the majority of Indians still live in rural areas, migration to cities is putting immense pressure on urban infrastructure systems and leading to high shares of urban dwellers being forced to reside in slums (about 20% in smaller cities and up to 40% in megacities) (Wagner, 2006). With urbanisation rates expected to rise further in the coming years, Indian cities are urgently searching for ways to tackle the growing challenges caused by increased pressure on energy, transport, water and waste management infrastructure.

The success of global climate mitigation efforts will depend on whether cities in India and other developing and emerging economies opt for carbon-intensive or low carbon solutions and whether collaboration with cities from industrialised countries, such as Germany, can foster a more climate-friendly transformation. To explore this, four examples of Indian-German urban cooperation have been selected for this study. Based on Yin's (2009) suggestions, the two major criteria for selecting the cases for analysis were (1) access to the required data (such as protagonists and experts for interviews, documents and opportunities for field observation) and (2) the likelihood of the cases to provide answers to the research questions. During initial research visits to German and Indian cities cases of urban cooperation were identified and assessed with regard to their stage of implementation and accessibility of information. From those cases offering sufficient research potential four cases were selected that exemplify the diversity of the institutional set-ups of transnational urban cooperation. Although not "ideal-types" (which hardly ever exist in reality), the four Indian-German urban partnership projects selected are largely representative of the distinct institutional forms highlighted in Andonova et al.'s (2009) typology of transnational governance networks (see section 3.2 of this study). The two partnership projects involving actors from Bremen and Pune are primarily *private* actor-driven (one NGO cooperation on decentralised wastewater treatment and one business-oriented partnership on public transport); the state-driven waste-to-energy project between Nashik and Hamburg is a largely *public* cooperation; and the clean energy partnership between Nagpur and Freiburg is run by ICLEI, a TMN that two of the typology's co-authors have classified as a *hybrid public-private* governance network in an earlier study (Betsill and Bulkeley, 2006).

4.6 Data Collection

This comparative case study analysis is based on a total of 56 interviews and document analysis.

In the early phase of the research project 20 exploratory interviews were conducted with city managers and decision makers, city partnership associations, city network and NGO representatives and researchers from Germany and India. Many of these interviews are not cited in the study as their primary aim was to identify potential case studies and gain initial insights into the transnational urban climate cooperation landscape and how these relationships function in practice. As part of the case study analysis an additional 36 semi-structured interviews were conducted with partnership actors and local stakeholders (16 interviews for the Pune-Bremen case studies, 11 for the Nashik-Hamburg case study and nine for the Nagpur-Freiburg case study). 31 of these interviews were arranged during research stays in the six cities involved between August 2012 and December 2013. An additional five telephone interviews were held in September and October 2014 to fill in information gaps and receive updates on the progress of the implementation (see an overview of the interviews in the appendix).

In order to protect the anonymity of the sources all interview data is not attributed to named individuals, rather to the affiliations they work for. However, it is important to note that their responses represent their personal opinions and not those of the named institutions.

The semi-structured interviews are a major source of information from which to assess the four cases according to the index system. The interview data was coded according to the indicators of the index system and the coded data was compiled in a database matrix.

The analysis of primary and secondary documentation served as an additional method of data collection to triangulate and supplement the data generated in the interviews. The documents, such as project master plans, feasibility studies, partnership contracts and agreements, partnership project reports, case studies, presentations and newspaper articles, were coded according to the same coding scheme as used for the interviews and the coded data was added to the database matrix, where the data was then scored.

5 Within Case Analysis

5.1 Case Study Pune – Bremen I: Developing Decentralised Wastewater Treatment Systems

5.1.1 Pune-Bremen History: A Longstanding Bottom-up Partnership in Sustainable Development

Non-state actors from Bremen and Pune have collaborated in environmental and social projects from as early as 1976. The partnership between the two cities was pioneered by Bremen citizen, Gunther Hilliges, who worked in Pune as head of the non-governmental organisation (NGO) Terre des Hommes Germany at that time. Hilliges had co-founded Terre des Hommes Germany a few years earlier and was responsible for setting up projects in Pune, home to the organisation's Indian office (Forum Städtesolidarität Bremen-Pune e.V., 2006). Terre des Hommes joined forces with local NGOs to develop projects supporting handicapped children, poverty alleviation and public health development in the city and Hilliges saw the potential to expand collaborative action between NGOs from Pune and Bremen.

One area of particular focus during the early years of the Pune-Bremen partnership was the introduction of micro-biogas plants in the rural areas around Pune. Hilliges initiated a cooperation between the NGOs "Maharashtra Arogya Mandal" (MAM) and "United Socio Economic *Development* and Research Programme" (UNDARP) from Pune and the "Bremen Overseas Research and Development Association" (BORDA), a non-profit research centre co-founded by Hilliges in 1977 to support Bremen's development cooperation activities. BORDA, MAM and UNDARP constructed more than 100 household-based biogas plants in the Pune area in the 1970s and early 1980s, funded by the Bremen city administration. The biogas plants replaced the traditional indoor combustion of cow dung in farmers' houses, treating the dung in a more efficient and smokeless way and converting it into gas for cooking and manure to be used as fertilizer on the farmers' fields. This project had several co-benefits. Not only were the households provided with the valuable end products of natural gas and fertiliser, the family members also suffered from less exposure to toxic smoke emissions. In particular the heath conditions of women, who usually are in charge of cooking and their children were thereby greatly improved.

Biogas remained a major topic in the Pune-Bremen cooperation. The partnership protagonists organised an international biogas conference series in Bremen in 1979 and 1984 and in Pune 1990. In their final declaration the conference participants highlighted the benefits of a worldwide diffusion of micro-biogas technology to improve ecological conditions, save energy and improve public health (Forum Städtesolidarität Bremen-Pune e.V., 2006).

The biogas project exemplifies how the Pune-Bremen partnership protagonists simultaneously promoted economic, social and environmental development from the outset of the city cooperation, many years before the concept of sustainable development was defined and promoted by the Brundtland Report in 1987 (United Nations World Commission on Environment and Development, 1987).

The transnational urban partnership between Pune and Bremen also stands out with regard to its informal institutional set-up. Particularly in the 1970s and 1980s, the Pune-Bremen cooperation was primarily based on private actor exchange and concrete NGO project development. An interviewee from the partnership association Forum Städtesolidarität Bremen-Pune points out that he and his supporters from Pune and Bremen deliberately decided to follow a bottom-up and civil society oriented approach in establishing bilateral relationships between the two cities. They were convinced that a partnership initiated and owned by local citizens and NGOs would yield better practical results than many of the city twinning partnerships that were established in a top-down manner by politicians after World War II. The interviewee refers to the example of the city partnership between Mumbai and Stuttgart that was promoted by German Chancellor Konrad Adenauer in 1953. According to the interviewee this partnership has never managed to involve local NGOs or deliver any practical results beyond the holding of art exhibitions and exchange visits by the two cities' mayors (Interview with Forum Städtesolidarität Bremen-Pune, 27-02-2013).

In the Pune-Bremen partnership the political level was bypassed for a long period and politicians were not involved in any activities during the first 20 years of the cooperation. To professionalise the partnership, broaden its scope beyond poverty alleviation and strengthen its connections with civil society, partnership associations were founded in both cities in 1980; the "Forum Städtesolidarität Bremen-Pune" in Bremen and the "Pune-Bremen City Solidarity Forum" in Pune. Hilliges convinced the NGOs from Pune with whom he had collaborated during his time as head of Terre des Hommes to join the Pune partnership forum; these included MAM and UNDARP, the NGOs that had run the partnership's biogas projects. In Bremen a total of one hundred local citizens, amongst them many

Terre des Hommes supporters, were involved in the foundation of the partnership association. The two forums were instrumental in developing the city partnership over the following years, together with the Association of the Friends of Germany (AFG) from Pune and the Bremen State Department for Development Cooperation (Landesamt für Entwicklungszusammenarbeit - LAFEZ). The AFG was founded in 1972 as a returnee association for former students and entrepreneurs who had spent time in Germany. The AFG works closely with the partnership association in Pune, running its office and organising events and exchanges between Pune and Bremen citizens. In the mid-1980s AFG member Vijay Mahajani also took over the chairmanship of the partnership association and became Bremen's main contact person in Pune until his death in 2010.

In 1979 the Senate of Bremen established the LAFEZ to coordinate and execute Bremen's development cooperation projects. Hilliges was appointed head of the LAFEZ and managed the partnership activities from this position until his retirement in 2005. When Hilliges started working for the LAFEZ in 1979, Bremen (a city state) was the first German state to engage in development cooperation, focusing on the areas of poverty alleviation, environmental protection and civil society empowerment. The interviewee from the Bremen partnership organisation Forum Städtesolidarität Bremen-Pune emphasises that the capacities of states and cities to conduct development cooperation work was neither recognized nor legally institutionalized in Germany at that time. He states that over the years cities have demonstrated that they can bring valuable expertise to North-South cooperation projects, so that today many German states involve cities in their engagement in development cooperation (Interview with Forum Städtesolidarität Bremen-Pune, 27-02-2013).

The LAFEZ adopted Hilliges' model of civil society oriented North-South cooperation and continued to develop NGO partnerships in India and South Africa, the two countries at the center of Bremen's development work. Over the years, the LAFEZ has established a network of NGOs with whom Bremen regularly cooperates and which can rely on continued funding from Bremen's city administration. In Pune, the LAFEZ has provided between 15,000 and 23,000 Euros in funding annually to a group of 12 to 14 NGOs since 1979 (Forum Städtesolidarität Bremen-Pune e.V., 2006). This funding is unique in that the LAFEZ has delegated the responsibility and administration of this fund to the partnership organisation AFG in Pune. The AFG meets with the local NGOs annually to reach a consensus on how the funds should be distributed. At the end of each year the AFG then sends a report to Bremen's city administration detailing which NGO projects were funded. Interviewees from Bremen emphasise that the distribution of the funds and financial reporting have always been carried out very well and that the trusting long-term cooperation with the AFG reduces the transaction costs of monitoring for Bremen's city administration (Interview with Forum Städtesolidarität Bremen-Pune and LAFEZ, 27-02-2013).

In addition to the biogas project and the support for NGOs from Pune the partnership also promoted exchange between schools, universities and local business. In 1983 universities from the cities of Bremen and Pune started collaborating by setting up joint research projects in the areas of environmental and development studies. Two years later the Pune and Bremen Chambers of Commerce signed a partnership agreement and carried out several joint trade fairs and projects (Bondre, 2005). In 1995 the Institute of International Business and Research (IIBR) was founded in Pune as a joint endeavour between the Mahratta Chamber of Commerce & Industries and the Pune Industrial Education Society, the LAFEZ, the Bremen Chamber of Commerce, Bremen University, and Leeds Metropolitan University. The LAFEZ supported the IIBR financially, the Bremen Chamber of Commerce served as consultant and Bremen University was responsible for developing the IIBR's academic structure and international orientation in teaching and applied research (Freie Hansestadt Bremen, 2005; Forum Städtesolidarität Bremen-Pune e.V., 2006).

The United Nations Conference on Environment and Development (the so-called "Earth Summit") in Rio de Janeiro in June 1992 was a major landmark for the Pune-Bremen partnership. As a result of the discussions about Local Agenda 21, local engagement in North-South cooperation was globally recognized for the first time. The end of the Cold War also helped pave the way for this new paradigm and the recognition of municipalities as crucial actors in international cooperation towards sustainable development (Interview with Forum Städtesolidarität Bremen-Pune, 27-02-2013). The Rio Earth Summit strongly affirmed the Pune-Bremen partnership protagonists' approach to simultaneously address economic, social and environmental development at the local level. It gave the partnership a boost and increasingly the cities' political actors became involved in the partnership.

In the aftermath of the Summit the LAFEZ funded several projects in Pune that were related to sustainable development: a prototype waste-to-energy plant at Pune University; energy plantations at the sewerage in cooperation with the Indian Institute of Education in Pune; environmental awareness projects for school children together with local NGO Arbutus; and for the first time, a joint project was conducted with the Pune Municipal Corporation, (Pune's city administration) to install nozzle-testing equipment in public bus depots to control and reduce bus exhaust emissions (Forum Städtesolidarität Bremen-Pune e.V., 2006).

Since 1995 both cities have intensified their efforts to develop and coordinate their Local Agenda 21 strategies. In March 1995 the "Pune-Bremen Collaborative Workshop on Sustainable Development & Regional Problems: Environment and

Economy" was held at the University of Pune. The conference participants, among them Hilliges and Mahajani, agreed to set up a Local Agenda 21 roundtable and develop a sustainable development action plan for Pune with citizen participation as its major focus.

During the international seminar on "Local Initiatives for Sustainable Development" held in November 1997 in Pune the partnership further widened its scope and invited political actors from both Pune and Bremen as well as officials from other Indian cities. The seminar's goal was to build the ground for an international network of European, African and South Asian cities supporting each other in the implementation of Local Agenda 21 processes. The seminar was jointly hosted by the AFG and the Pune Municipal Corporation and representatives of 17 Indian cities participated in the event. Pune's Mayor Vandana Chavan officially invited Bremen's Mayor Henning Scherf and Hilliges to join the seminar (Forum Städtesolidarität Bremen-Pune e.V., 2006). More than 20 years after the first contact between Pune and Bremen this was the first ever visit by a senior political representative from Bremen to Pune (Interview with Forum Städtesolidarität Bremen-Pune, 27-02-2013). As part of the visit, the Chief Minister of Maharashtra, Shri Manohar Joshi, and Scherf inaugurated the Pune-Bremen Friendship Square on November 11th 1997 (Bondre, 2005).

In 1998 Bremen and Pune (together with Pune's neighbouring city Pimpri Chinchwad) signed the first partnership Memorandum of Understandung (MoU) and agreed to establish a joint International Office Agenda 21 (IOA 21) in Pune to mainstream sustainable development in the three cities and institutionalize the cooperation. The IOA 21's main tasks were to promote and coordinate joint partnership projects, exchange visits and conferences, plus support Pune and Pimpri Chinchwad in their international networking to expand Local Agenda 21 processes in South Asia. The partner cities agreed to split the financing of one permanent IOA 21 employee and office equipment. Over the following years the IO A21 organised and supported several partnership events, including an international conference on city-level sustainable development held at Pune University in March 2001 involving about 250 participants from six South Asian countries. The conference findings were presented at the Second International Congress on "Business and Municipality – Creating Better Cities Together" two months later in Bremen (Freie Hansestadt Bremen, 2005).

In 2003 Bremen, Pune and Pimpri Chinchwad agreed to continue their support for the IOA 21 and they extended the partnership MoU for another three years. In the renewed MoU the partner cities decided to establish an Agenda 21 stakeholder roundtable in Pune. They highlighted joint research and education projects as well as technology transfer as key areas of cooperation, particularly in the areas of public transport, waste management and drinking water (Forum Städtesolidarität Bremen-Pune e.V., 2006; Hilliges, 2006). The partner cities focused in particular on two partnership projects during the period of the second MoU; the introduction of decentralized wastewater treatment systems (DEWATS) in Pune and the transfer of the Bremen tramway system to Pune and Pimpri Chinchwad (both projects have been selected as case studies for this thesis, see sections 5.1 and 5.2). Since 2006 the partnership MoU has not been updated and partnership activities have slowed down. One reason for this slowdown was the retirement of Hilliges, the initiator and main driving force behind the Pune-Bremen partnership, in 2005. Hilliges remained in close contact with the LAFEZ employees, he was elected as chairman of the Bremen partnership association "Forum Städtesolidarität Bremen-Pune", and he continued to travel to Pune regularly at his own expense. It was however not possible for him to maintain the relations with the Pune Municipal Corporation that were established at the start of the Local Agenda 21 activities in the mid-1990s. As the cooperation between Pune and Bremen was not formalized through a twinning partnership agreement and Hilliges was not officially representing the Bremen city administration anymore, it proved difficult to gain access to the regularly changing political leadership in Pune and to convince them of the benefits of the partnership with Bremen. A partnership actor from Bremen highlights that frequent changes to the city's political leaders have been a key barrier to maintaining contact with Pune Municipal Corporation (Interview with Forum Städtesolidarität Bremen-Pune, 27-02-2013). During two exchange visits to Bremen and Pune in 2010 Hilliges attempted once again to prepare a new partnership MoU together with a representative of the NGO Arbutus who has become Bremen's main contact person in Pune after Mahajani passed away. However, the contact with the Pune Municipal Corporation and the IOA 21 could not be re-established. By the time of data collection for this study (October 2012 to January 2014), all other interviewed partnership actors from Bremen and Pune had also lost touch with the IOA 21. A Pune Municipal Corporation representative explains that the office is still officially there, but it is only used for managing Pune's official twinning partnerships with Okayama (Japan) and San José (USA), rather than the cooperation with Bremen (Interview with PMC, 24-09-2013).

Prior to Hilliges' visit in 2010, already one attempt by Bremen and Pune citizens to revive the partnership of non-state actors had failed. In 2008 and 2009 a German journalist living in Pune, and the deputy chairwomen of the Bremen partnership association Forum Städtesolidarität Bremen-Pune, tried to establish a new forum of exchange between NGOs from Pune and Bremen. They organised meetings in Pune where several renowned local NGOs (amongst others Parisar, an influential NGO in Pune working on transport issues) expressed their interest in intensifying the cooperation with NGOs from Bremen. However, this initiative never came into existence, due to a lack of feedback and engagement from the Bremen side, as the journalist involved in this initiative points out (Interview with journalist from Pune, 18-09-2013).

Despite these setbacks the LAFEZ continues to financially support the group of 12 to 14 NGOs in Pune as well as the Bremen partnership association "Forum Städtesolidarität Bremen-Pune" which regularly organises seminars and lectures about India and Pune (Interview with LAFEZ, 27-02-2013; Forum Städtesolidarität Bremen-Pune e.V., 2011). In 2012, Hilliges and the Pune NGO Arbutus started a new initiative in the field of education to revitalize the Pune-Bremen partnership. Arbutus is active in the area of environmental education and plans to set up exchange projects among schools, teacher training colleges and universities in Pune and Bremen. During mutual visits by Pune and Bremen delegations in August and September 2013 Annette Lang, the coordinator of international relations at Bremen University, stated that the biology, history and language departments of the Bremen University were interested in collaborating with the respective university departments of Pune University. The representative of Arbutus voiced confidence that this initiative will give new impetus to the Pune-Bremen cooperation (Interview with NGO Arbutus, 25-09-2013). A first step towards more academic cooperation has been made in 2014. Twenty-two students from Pune registered for the Hochschule Bremen's 2014 International Summer School for German Language, Culture and Business (Kabbert, 2014).

5.1.2 Partnership Project DEWATS: Content and Process

During the second city partnership MoU negotiations in 2003 the Pune and Bremen partnership protagonists agreed to jointly develop solutions for waste and wastewater as one major field of cooperation. The partnership actors decided to test the applicability of Decentralized Wastewater Treatment Systems (DEWATS) in Pune as an alternative solution to the city's overburdened sewage system.

The NGO BORDA from Bremen, which had already been involved in the Pune-Bremen biogas projects in the 1970s and 1980s, joined the city partnership once again. Since 1994 BORDA has developed DEWATS as a decentralized sanitation solution for developing countries and emerging economies of the Global South. The DEWATS concept is based on a modular design of several sequences of natural wastewater treatment, meaning it can be adjusted to local requirements and different volumes of daily wastewater inflows of between one and 1000m^3[45].

---

45 http://www.sswm.info/category/implementation-tools/wastewater-treatment/hardware/site-stor age-and-treatments/biogas-settl (19-02-2016)

BORDA promotes DEWATS as a long-lasting, frugal wastewater treatment technology for households or small to medium-sized industries. DEWATS plants can be constructed with locally available material. They meet environmental regulatory standards and reduce water pollution by up to 90% compared to non-treatment. As DEWATS plants are based on ecological waste treatment processes, they require no energy inputs and require very little maintenance. The only end products are service water, fertilizing sludge and – depending on the selection of modules – biogas, all of which are highly demanded resources in many developing countries (Bremen Overseas Research and Development Association, 2010).

BORDA had already implemented DEWATS plants in other Asian countries, such as China and Indonesia before it started focusing on India. In 2001 BORDA successfully applied for a grant from the European Commission to send one permanent employee, Pedro Kraemer, to Bangalore. In the following years Kraemer set up the Consortium for DEWATS Dissemination (CDD) Society, a local training centre and contact point for BORDA's engagement in India. BORDA and the CDD built several DEWATS plants in Southern India (Kerala and Pondicherry) and Delhi before they started working in Pune (Interview with NGO BORDA, 11-09-2014).

The DEWATS initiative in Pune began with two partnership events in Pune and Bremen in 2003 where Pune representatives impressed their partners from Bremen by voicing strong interest in developing DEWATS in Pune. As a result, the decision was taken to start building several pilot DEWATS plants in Pune and assess the scope for a wider dissemination of the technology in the city (Interview with DEWATS project consultant, 08-10-2013). The first plants were built in Hadapsar, a former suburb of Pune. Hilliges and Gujar, head of the local Sane Guruji Ayurveda Hospital and chairman of the NGO Maharashtra Arogya Mandal (MAM) decided together with BORDA and the German NGO Arbeiterwohlfahrt (AWO) that the Ayurveda hospital would be an adequate facility in which to build the first DEWATs demonstration plants. Both BORDA and AWO had already collaborated with Gujar on prior projects and Hilliges and Gujar knew each other personally (Interviews with NGO MAM, 19-09-2013 and BORDA, 11-09-2014). In 2004 and 2005 BORDA erected two DEWATS plants for the hospital in Hadapsar (capacity: $35m^3$/day each), plus one smaller plant (capacity: $7.5m^3$/day) in a MAM-run hostel for tribal students in Narodi, a village outside of Pune (Gujar, 2010).

The projects were jointly planned and implemented by BORDA employee Andreas Schmidt, local engineer Nitin V. Patil and Gujar who supervised the project in Pune. Due to severe water scarcity in Pune they decided on a DEWATS system that only cleans the wastewater, opting against additional biogas production as this would have been too water intensive. The plants consist of four modules. The primary treatment and sedimentation in a settler is followed by a secondary anaerobic treatment in a baffled upstream reactor. Then tertiary aerobic and anaerobic treatment is conducted in planted gravel filters and finally in ponds (Gujar, 2010). The resulting service water is used to irrigate medical plants in the Aryurvedic hospital's garden. Gujar opted against the utilisation of the water in the hospital latrines. He was concerned that the hospital guests may have been afraid of infections being transmitted via contact with the water and this could have adversely affected the hospital's guest numbers (Interview with NGO MAM, 19-09-2013).

During the implementation phase, a German engineer and expert in decentralised wastewater and biogas technology joined the project team. The engineer had been approached by Kraemer to apply for the position of "Advisor on Dissemination of Community Based Sanitation and Decentralized Wastewater Treatment Projects" on the Centre for International Migration and Development (CIM) placement program. This program supports global labour mobility and is jointly run by the German government agencies, the GIZ and the German Federal Employment Agency (Bundesagentur für Arbeit – BA). The advisor's placement in Pune was approved by the CIM due to BORDA's experiences and success in Asia and because the CIM saw the creation of the partnership's International Office Agenda 21 and the concept to develop DEWATS in Pune to be promising (Interview with project avisor, 08-10-2013). The advisor was sent to Pune to help finalise the pilot plants, build the ground for additional DEWATS projects and assess the scope for a wider application of the technology in the city. He worked in Pune at the IOA 21 for a total 18 months between 2005 and 2007. During this time the three DEWATS plants in Hadapsar and Narodi were commissioned and the advisor supervised the construction of two additional DEWATS plants that BORDA built at the Matru Mandir orphanage in Pune's neighbouring town Devrukh, plus one plant for Lawkim Ldt., a private investor from Thane.

The partnership actors involved did not achieve their goal of facilitating a larger scale dissemination of DEWATS in Pune. The advisor recalls a failed attempt to build a DEWATS plant in a local school in Pune funded by a private investor. According to this interviewee the investor never fully understood the technology and turned out to be unreliable so the project was never realised (Interview with DEWATS project advisor, 08-10-2013).

Over the course of the advisor's stay in Pune support from the Pune Municipal Corporation (PMC) also began to wane (ibid.). Initially, the PMC showed strong interest in DEWATS and engaged in setting up its own pilot DEWATS project in a new residential area consisting of 1000 houses for former slum dwellers in Hadapsar (Freie Hansestadt Bremen, 2005). But this plant remained the only

DEWATS facility in which the PMC was involved. It only functioned for a short time after completion before it eventually failed. After working in Pune for 18 months the advisor did not see any prospects for the realisation of further DEWATS plants so he ended his engagement in Pune and left the city in 2007 (Interview with DEWATS project advisor, 08-10-2013).

5.1.3 Partnership Project DEWATS: Outcomes

In this comparative case study analysis, success in transnational urban partnership projects has been defined and assessed according to seven indicators measuring the project implementation and the quality of the wider project outcome (see section 4.4).

The remarkable feature of the DEWATS project evaluation is that it achieves a medium score on most of the success indicators. The partnership initiative does not positively excel in any of these indictors even though it had one clear weakness: the lack of mutuality with regard to benefits and learning (see table 2).

Success Indicators	Project Evaluation: DEWATS		
	Pune	Bremen	Total
Implementation Achievements	/	/	+-
Budget and Schedule Performance	/	/	+-
Local Capacity Building	+-	+-	+-
Benefits for Target Group	/	/	+-
Impact	/	/	+-
Mutuality	/	/	--
Post-project Sustainability	/	/	+-

Table 2: Partnership Project DEWATS: Findings from success index

Implementation Achievements

The Pune-Bremen partnership succeeded in building five pilot DEWATS plants in Pune and surrounding cities as a result of cooperation between NGOs (the DEWATS plant built for the private investor Lawkim Ldt. is excluded from this analysis due to lack of sufficient data). The three DEWATS plants built under the BORDA-MAM cooperation (two at the Sane Guruji Hospitalin Hadapsar and one at the student hostel in Narodi) were the first DEWATS plants ever built in the Indian state of Maharashtra. They have been functioning well since they started operating in 2005, treating 100% of the wastewater produced in these facilities. The plants also require very little maintenance: "To operate the treatment unit, skills of gardener or a caretaker are enough." (Gujar, 2010). The only tasks are the removal of the sludge every 12 to 24 months, occasional checking of the development of algae and regular weeding of the plants. The operation and maintenance costs of the DEWATS plants are thus very low. However, the interviewee from MAM states that the construction costs of an average 100 Rupees per litre of capacity are still too high for the Indian context so currently plants cannot be built without external financial support (Interview with NGO MAM, 19-09-2013). In addition the two plants at the Matru Mandir hospital in Devrukh were also successfully built and are operational (Interview with DEWATS project advisor, 15-09-2014).

Two additional plants, involving the Pune Municipal Corporation and a private investor respectively, failed. The DEWATS plant in the new residential area in Hadapsar that was built by Pune Municipal Corporation was due to serve as a demonstration project for other poor areas in Pune (Freie Hansestadt Bremen, 2005). The plant was constructed between 2005 and 2007 in cooperation with BORDA. However according to the interviewee from MAM the plant has never worked properly and by 2013 was no longer in operation (Interview with NGO MAM, 19-09-2013). The DEWATS plant that the private investor wanted to build for a local school was not even fully constructed (Interview with DEWATS project advisor, 15-09-2014).

Also the planned city-wide dissemination of the technology was never realised. Since the DEWATS project advisor left Pune in 2007 no additional DEWATS plants have been built in the city. An interviewee concludes that the DEWATS partnership project started ambitiously, but in the end it did not have much effect in Pune and the problem of insufficient wastewater treatment remains pressing (Interview with journalist from Pune, 18-09-2013).

Budget and Schedule Performance

The five NGO-led DEWATS projects in Hadapsar, Narodi and Devrukh were in line with the calculated budget. The DEWATS project advisor states that the financial budget of the NGO-led DEWATS projects had to be adhered to as there was no additional funding available (Interview with DEWATS project advisor, 08-10-2013). According to a project report, in particular the two plants in Devrukh were in a "good budget situation" (Klatte, 2005, 9). The NGO-led DEWATS projects were also largely constructed within the projected time schedules. Only the DEWATS plants built by BORDA and MAM in Hadapsar were delayed slightly when unexpected rock formations were found in the ground and the construction work had to be stopped until the rainy season was over. But according to the project advisor these delays remained limited, mainly due to the fact that Gujar took ownership and responsibility for the project (Interview with DEWATS project advisor, 08-10-2013). The advisor points out that the two DEWATS plants in Devrukh that were built by BORDA and Matru Mandir did not exceed the budget or time schedules (Interview with DEWATS project advisor, 15-09-2014).

According to the project advisor, the only project that faced major delays was the PMC-funded plant. The project schedule had to be extended several times, meaning that the plant was only commissioned in 2007 (ibid.).

Local Capacity Building

The assessment of the third success indicator, local capacity building as part of the partnership project, is likewise mixed. On the one hand, the DEWATS projects led to organisational learning for the participating NGOs, BORDA, MAM and Matru Mandir. On the other hand the projects had no effect on administrative processes or institutional reforms in Pune and Bremen.

The NGO BORDA from Bremen utilized the demonstration plants in Pune to explore the scope for DEWATS in India. BORDA observed and studied the operating plants for several years and improved the DEWATS technology based on the reviews. The German project advisor points to the example of the baffled reactor sequence which BORDA has downsized as a result of experiences from Pune (Interview with DEWATS project advisor, 08-10-2013). Moreover, BORDA drew on the know-how from the first projects in Pune and other Indian cities to develop its knowledge and training centre CDD in Bangalore. The NGO MAM and the Sane Guruji Hospital also serve as a local information centre for DEWATS in Pune and Maharahstra, albeit on a smaller scale than the CDD. The DEWATS plants in Hadapsar are the oldest operating DEWATS facilities in the state of Maharashtra and they have served as a learning model for interested visitors. BORDA has repeatedly visited the plants together with international students to showcase the technology. Once even the Maharashtra State Minister inspected the DEWATS plants in Hadapsar and asked MAM for the technical drawings to provide them to his department engineers for replication (Interview with NGO MAM, 19-09-2013). Today the hospital operates and maintains the DEWATS plants independently and MAM has compiled an information leaflet about the DEWATS technology and demonstration plants (Gujar, 2010).

However, the goal to build enough capacity at the Pune city administration to make DEWATS part of the city's sanitation portfolio was not achieved. The IOA 21 and BORDA conducted several workshops in Bremen, Pune and Bangalore to showcase the technology and motivate and train Pune city administration staff and other stakeholders to engage in DEWATS projects. But this had no real effect as the project advisor explains. He states that these workshops were not sufficient to make the city administration staff understand the technology and scope of DEWATS plants well enough. He recalls several occasions when the PMC engineers suggested locations for DEWATS plants that were far too large in their scope. Throughout his stay in Pune the project advisor had the general impression that the city administration remained rather ignorant and sceptical of the small-scale, decentralised sanitation option (Interview with DEWATS project advisor, 08-10-2013).

With the exception of BORDA's organisational learning (see above), there was also no capacity building in the city of Bremen.

Benefits for Target Groups

The impact of the project on its target groups was also mixed. The residents of the facilities where the well-functioning DEWATS plants were built have clearly benefited from the partnership initiative as the pilot plants have enhanced the sanitary conditions for the people living and working at these sites. At the three DEWATS plants that I visited in September 2013 (the MAM-run plants in Hadapsar and Narodi) the people praised the efficiency, low maintenance and valuable end projects of the plants. In particular the DEWATS plants in Hadapsar met the hospital's needs: the plants treat all wastewater and provide the Ayurvedic clinic with service water and fertilizing sludge which is reused for the hospital's medical garden. The installation of DEWATS plants has also strongly improved the respective local environmental and health conditions at the hospital, as prior to the project's implementation the wastewater had been pumped onto a nearby field where it was a breeding ground for malaria-carrying mosquitos (Interview with DEWATS project advisor, 08-10-2013).

The interviewee from MAM is very content with the results of the DEWATS plants as the maintenance costs are very low, no electricity is required to run the plants, and the remaining sludge is reused as manure for the hospital's medical plantations (Interview with NGO MAM, 19-09-2013).

For the majority of the citizens in Pune however the DEWATS partnership project brought little improvements as the PMC plant for former slum people and the city-wide dissemination of the DEWATS technology failed. An interviewee explains that there has been no diffusion of the DEWATS concept in the city and the demonstration plants have overall had limited impact in Pune (Interview with journalist from Pune, 18-09-2013).

Impact beyond the Partnership Project

While the impact of the DEWATS demonstration plants in the city of Pune remained limited, the partnership project still had some wider effects outside of Pune. BORDA and the CDD have established a leading international conference series on DEWATS based on the experiences of the first pilot plants in Pune and other Indian and Asian cities. The conference series was initiated at the first International Workshop on Decentralized Wastewater Treatment Systems and Community-Based Sanitation in Pune in 2003 (Interview with Forum Städtesolidarität Bremen-Pune, 27-02-2013). The following DEWATS conferences were organised in cooperation with the International Water Association (IWA) in Hanoi, Vietnam (2008), Surabaya, Indonesia (2010) and Manila, Philippines (2011), before in 2012 the series returned to India and was hosted by the city of Nagpur. The Nagpur conference was visited by more than 200 city officials, businessmen and researchers from all around the world and a core finding from the event was that DEWATS has developed from a niche application to a wide-spread solution for urban sanitation in Asia. This was demonstrated to the conference participants on a visit to a DEWATS plant that was recently built as part of a local slum development project in Nagpur.

(Perceived) Partnership Mutuality

As for the success indicator of measuring partnership mutuality, the DEWATS project contains many elements of a one-sided North-South transfer of lesson-drawing and benefits, even if the NGO BORDA from Bremen did learn something from the initiative (see section on Local Capacity Building above). The technical know-how that was required for setting up the DEWATS plants was brought to Pune by BORDA and the German project advisor. The five NGO-run DEWATS plants were also funded by a German NGO, the Arbeiterwohlfahrt, and the project advisor was paid by the German government's CIM program (Interview with DEWATS project advisor, 08-10-2013). Moreover, there was no direct technical knowledge transfer back to the city of Bremen as part of this partnership project, as the whole city has been already connected to a centralized wastewater treatment system and there is not much scope for DEWATS.

The one-sided transfer of funds from Bremen to Pune was standard in the many activities that were initiated as part of the city partnership. The LAFEZ has continued to financially support joint projects and personal exchange, as the financial capacity from the side of Pune to conduct partnership projects has been limited. While in the early years many decision makers in Bremen were rather sceptical about the one-sided partnership funding, the LAFEZ's development cooperation work in Pune and other cities of the Global South has gained more support in recent years by Bremen's politicians (Interview with Forum Städtesolidarität Bremen-Pune, 27-02-2013).

In fact, the LAFEZ budget for development cooperation was the only area explicitly excluded from any cuts in the coalition contract of the new Bremen state government in 2011 (Sozialdemokratische Partei Deutschlands, Landesorganization Bremen Bündnis 90/Die Grünen Landesverband Bremen, 2011, 118). The interviewees from the LAFEZ and the Forum Städtesolidarität Bremen-Pune emphasise that this is remarkable, as the city is facing a severe budgetary crisis and many other departments had to accept large budget cuts (Interview with Forum Städtesolidarität Bremen-Pune and LAFEZ, 27-02-2013).

Post-project Sustainability of the City Partnership

While the Bremen state department LAFEZ continues to support the Pune-Bremen partnership, the Pune Municipal Corporation withdrew from the cooperation after the DEWATS and tramway projects. The two interviewees from PMC were not aware of any partnership activities after these two projects. They state that the partnership was more active during the years around the partnership MoU in 1998, but then activities slowed down due to changes in the political leadership, lack of personal exchange and lack of concrete projects. They explain that setting up policies such as MoUs is easy but that their implementation is the crucial challenge, a challenge that they consider the Pune-Bremen cooperation to have failed to overcome. When asked if the PMC has any future plans as part of the partnership they said that currently no concrete projects are prepared (Interviews with PMC, 29-11-2012; 24-09-2013).

The interviewees from LAFEZ and Forum Städtesolidarität Bremen-Pune confirm that they have lost contact with the PMC and the IOA 21 in Pune and that the partnership activities with Pune have been reduced in recent years, since the

PMC's support has waned and their long-standing partner and contact person in Pune, Mahajani, passed away in 2010. The LAFEZ has continued to offer financial support of currently around 18,000 Euros to the group of 12-14 NGOs in Pune and has also organised several other partnership activities such as exchanges between local schools, museums and hospitals from Pune and Bremen. But the partnership has not again reached the intensity it exhibited during the periods of the partnership MoUs (1998-2006). The partnership protagonists hope for a renewed impetus in the coming years from a revival of the cooperation in the area of education (Interviews with Forum Städtesolidarität Bremen-Pune and LAFEZ, 27-02-2013 and NGO Arbutus, 25-09-2013).

5.1.4 Partnership Project DEWATS: Explanatory Factors

In view of the success factors outlined above, the following section explores explanations for the project's performance in these areas. It tests four hypotheses on success conditions for transnational urban partnership projects derived from existing literature as well as my own research: (1) the existence of an adequate strategy for knowledge exchange, (2) the linkages between partnership projects and state institutions, (3) the existence of engaged and well-connected local partnership entrepreneurs, and (4) the development of partnership social capital (see chapters 3 and 4.3 for more detailed information on how the hypotheses and the index system were developed).

5.1.4.1 Knowledge Exchange Strategy

The first hypothesis refers to the process of knowledge exchange among partnership cities, arguing that a partnership project is more likely to succeed if it follows a well-prepared and locally-tailored strategy of cooperation.

In the Pune-Bremen DEWATS project this hypothesis was partly confirmed, as table 3 indicates. This partnership project scores highly in the indicators of "Prior Program Evaluation" and "Continuity of Interaction". The scorings of the indicators "External Moderator/Facilitator" and "Intrinsic Interests" are however low.

Indicator: Knowledge Exchange Strategy	Project Evaluation: DEWATS		
	Pune	Bremen	Total
Prior Program Evaluation	/	/	++
Prior Context Evaluation	/	/	+-
Systematic Cooperation	/	/	+-
Continuity of Interaction	/	/	++
External Moderator/Facilitator	/	/	--
Policy Window	/	/	+-
Intrinsic Interests	+-	--	+--

Table 3: Partnership Project DEWATS: Findings from Index Knowledge Exchange Strategy

Prior Program and Context Evaluation

In the DEWATS partnership project the coordinating NGO BORDA from Bremen drew on extensive experiences and technical expertise from prior projects. Since BORDA's foundation in 1977 the NGO has specialized in small-scale and low-maintenance eco-technologies in renewable energies (biogas, hydropower) and waste and water management, tailored to the context conditions of developing countries (Freie Hansestadt Bremen, 2005). BORDA has established itself as one of the leading DEWATS providers in Asia, South Africa and Latin America where it runs regional offices. The interviewee from Forum Städtesolidarität Bremen-Pune points out that BORDA systematically developed the DEWATS technology, with first applications in China, Indonesia and South India and that these facilities served as reference models for the plants in Pune (Interview with Forum Städtesolidarität Bremen-Pune, 27-02-2013). The German project consultant confirms that BORDA particularly drew on the experiences of its DEWATS plants in Bangalore and Indonesia where Schmidt had worked before he started supervising the project in Pune (Interview with DEWATS project advisor, 08-10-2013). An example of the transfer of knowledge from prior projects is the baffled reactor module, an innovative septic tank with upstream water treatment, which was integrated in the plants in Pune after it was successfully tested in BORDA's earlier projects (ibid.).

According to the project advisor an in-depth context evaluation is generally a prerequisite for setting up any DEWATS plant, as the application of the modules always has to be adjusted to the respective local conditions. He states that BORDA and their local partner NGOs MAM and Matru Mandir conducted extended evaluations of the specific context conditions for the DEWATS pilot sites. The plans for the DEWATS plants at the Hadapsar hospital were modified several times after the BORDA and MAM had analysed the local conditions. Due to Pune's dry climate and frequent water scarcity the project partners decided against the water-intensive biogas production and instead were able to use the treated water for the hospital's medical garden. Furthermore the initial location for the DEWATS facility at the hospital was changed as it turned out to be too small (Interview with DEWATS project advisor, 08-10-2013).

Only in the case of the PMC plant was an extensive context evaluation not conducted. The project advisor explains that the site was selected in a rather ad hoc manner and without in-depth considerations about the later operational implications. He also points out that he and his colleagues from BORDA generally overestimated the political support for DEWATS in Pune. Particularly the influence of the International Office Agenda 21 in Pune as well as the commitment of the Pune Municipal Corporation to disseminating the DEWATS technology across the city turned out to be much lower than expected. The PMC representatives had voiced strong interest and commitment to decentralized wastewater treatment at the beginning of the project. But when the project advisor arrived in Pune he realised that the PMC staff responsible had still not fully understood the process and scope of the DEWATS technology. He also realised that the PMC officials were actually more interested in setting up a large-scale centralized wastewater treatment system rather than the decentralized and small-scale DEWATS plants (Interviews with project advisor, 08-10-2013; 15-09-2013).

Systematic Cooperation, Continuity of Interaction and External Moderator/Facilitator

The mixed results in implementing DEWATS in Pune – the successful construction of DEWATS pilot plants through NGO cooperation and the failed collaboration with the PMC to disseminate the technology on a larger scale – can also be explained by the fact that the NGO-run plants were implemented in a more systematic manner than the PMC-led plant. The NGO-led DEWATS plants were the product of joint collaborations between NGOs from both cities. These plants were constructed in locations where stakeholders had clearly voiced a need for decentralized wastewater facilities. BORDA and the partner NGOs MAM and Matru Mandir developed master plans, modified these according to the on-site analysis of the local conditions and then constructed the plants together with local construction workers (Interview with NGO MAM, 19-09-2013). The cooperation between BORDA and the PMC to facilitate a wider dissemination of DEWATS in

Pune was less systematic. The project advisor explains that BORDA had expected the PMC to provide a comprehensive mapping of possible locations for the DEWATS plants. Based on this overview the best locations were to be selected and then both partners were to acquire funding for them. But the mapping was never conducted and PMC employees kept bringing the project advisor to locations that were unsuitable for the plants. He adds that the plan to disseminate DEWATS in Pune did not follow a clear problem-based approach as there was little intrinsic motivation within the city administration to establish the cooperation and no groundwork had been initiated (Interview with DEWATS project advisor, 08-10-2013).

The project advisor considers this to be a challenge that is widespread in development cooperation, where projects are often initiated because funds are available and because of the personal prestige and the opportunities to travel enjoyed by the partners, rather than because of actual need. According to his account in 2005 the PMC was no longer serious about pursuing the Local Agenda 21 process which had been introduced by the Pune-Bremen partnership MoUs in 1998 and 2003. The project advisor highlights that during his time in Pune between 2005 and 2007 there was no systematic development of a sustainability strategy, no systematic stakeholder involvement and no competent staff in the city administration that could have put sustainable development on the political agenda (Interviews with project advisor, 08-10-2013 and 15-09-2014).

Another key challenge that transnational urban North-South cooperation often faces is maintaining continued interaction between the partnership actors throughout the whole project implementation process. Project partners often struggle to maintain contact due to the geographical distance and different communication cultures. In the case of the DEWATS partnership projects the fact that BORDA had esta-blished a regional office in Bangalore facilitated regular personal exchange. BORDA employees Schmidt and Kraemer regularly visited the project sites in person to observe the project's progress. When the German project advisor started working at the IOA 21 in 2005, BORDA even had a permanent contact person in Pune to supervise the construction of the pilot plants. The advisor stayed for a total of 18 months and during this time the five NGO-led DEWATS plants in Hadapsar, Narodi and Devrukh were successfully built. But despite his permanent presence in Pune he did not see any further progress in the dissemination of DEWATS technology and he prematurely terminated his stay in 2007. In retrospect the advisor believes that his stay may not even have been long enough to build enough local contacts and capacity to make a more fundamental impact (Interview with project advisor, 08-10-2013).

A critical question in the development and implementation of city-to-city projects is whether it is beneficial to rely on the expertise, contacts and financial means of external partnership moderators, such as government ministries, development agencies or transnational city networks. Since its initiation in the 1970s the Pune-Bremen partnership has primarily focused on facilitating direct exchange between local citizens, NGOs, educational institutions and business without external support. Two members of the partnership association, AFG, explain that direct exchange without outside help usually reduces bureaucratic efforts and saves time (Interview with AFG, 26-11-2012).

As in most of the other partnership projects between Pune and Bremen, the DEWATS initiative was set up without the facilitation of external moderators. The funds provided by the AWO and the German government's CIM program were the only external contributions to the project.

While the direct NGO exchange in the project led to the realisation of small-scale demonstration plants in Pune, the lack of external involvement may have been one reason why the dissemination of the DEWATS technology in the city failed. Even if there has been a transfer of policy-making competences to develop sanitation strategies to city governments in India, urban local bodies often lack the resources, capacity and experiences to independently introduce innovative technologies. Andreas Ullrich, the former director of BORDA, also calls for more external support to scale up decentralized sanitation concepts, as BORDA alone lacks the capacity to disseminate its technologies more widely (Interview in taz, die Tageszeitung, 24-08-2009).

Policy Windows and Intrinsic Interests

As for the local political conditions the existence of policy windows and their utilisation by project advocates has been highlighted as a critical factor for project implementation. According to Kingdon (2011), policy windows occur when the problem stream, policy stream and political stream join. In the DEWATS partnership project this condition was only partly met. The problem – the lack of sufficient sanitation facilities in many areas of the city of Pune – is pressing. In Pune there is a high demand for alternative sewage treatment as the regular overflowing of sewage already causes serious environmental and health problems (Gaikwad, 2013). The sewage system is under additional pressure as many existing residential areas are not yet connected to Pune's sewage system and in the last years several new suburbs have been under construction. With regard to the policy stream the DEWATS technology can be regarded as an adequate technical solution to complement and relieve the pressure on the existing sewage treatment system. However, DEWATS plants require space which is limited in the densely-populated centre of Pune and they entail high initial investment costs which impede their dissemination or make them reliant on external funding. An assessment of the political stream provides similarly mixed results. On the one hand the impetus to initiate and coordinate Local Agenda 21 processes in Pune and Bremen and set up the partnership MoUs and the IOA 21 came from the United Nations Earth Summit on Sustainable Development in Rio de Janeiro 1992 (Interview with Forum Städtesolidarität Bremen-Pune, 27-02-2013). The international political background helped facilitate the DEWATS initiative. However, initial local political interest and support for the DEWATS technology was waning which meant that only the NGO-led plants were successfully realised. An interviewee states that although there is a pressing need for environmental improvements there is too little commitment by political decision makers to facilitate any environmental policies in Pune (Interview with journalist from Pune, 18-09-2013).

Another projected success factor for implementing urban low carbon initiatives is to closely align joint projects with both project partners' intrinsic interests. A key strategy for climate projects in India is the focus and promotion of their economic, social and environmental co-benefits. In Pune, the DEWATS plants were promoted as an energy saving, low maintenance technology to clean water and improve health conditions, rather than a low carbon innovation (International Office Agenda 21, n.d.). The DEWATS brochure for visitors of the MAM-run DEWATS plants in Hadapsar praises the efficient and pollution-free treatment of water: "This system helps to minimise water pollution ensuring that water is used economically and is reused to the maximum possible extent after treatment. The treated water can be reused for irrigation, groundwater recharge." (Gujar, 2010). The project advisor also used to highlight the co-benefits of the simultaneous wastewater treatment, water recycling and health improvements when he presented the project in Pune, as these topics were of more concern to the local people than the climate-friendly effects of the DEWATS technology (Interview with DEWATS project advisor, 08-10-2013).

Despite these efforts, the city administration and political decision makers from Pune remained sceptical that the city-wide dissemination of DEWATS would serve their interests. The project advisor states that the PMC and local politicians generally preferred more prestigious larger-scale projects, rather than the small-scale DEWATS facilities (Interview with DEWATS project advisor, 08-10-2013).

According to the project advisor, the Bremen and Pune partners' priorities in the project differed substantially. He presumes that the PMC wanted the German partners to bring in the project funding and be responsible for the project's execution, while the actors from Bremen had expected the PMC to engage much more pro-actively in the dissemination of DEWATS in Pune (Interview with DEWATS project advisor, 11-09-2014).

In contrast to the local officials from Pune, BORDA and the LAFEZ were driven by the idea of establishing DEWATS as a viable alternative to centralized wastewater treatment across the city of Pune. BORDA was even willing to offer the know-how for building DEWATS plants free of charge to all project partners from Pune, but only the NGOs MAN and Mantru Mandir were fully convinced that DEWATS provides an adequate solution to their sanitation problems (Interview with DEWATS project advisor, 08-10-2013).

The promotion of the DEWATS initiative and their co-benefits in the city of Bremen was not part of the project strategy. There were no plans to build any DEWATS facilities in Bremen where all households and industries already have access to the existing centralized wastewater system. As a result, Bremen pursued no intrinsic interests in this partnership project.

5.1.4.2 Linkages between the Partnership Project and State Institutions

The second hypothesis argues that the more a transnational urban partnership project is institutionalized into the state system, the more likely the project is to succeed. The case study findings reveal that of the four cases analysed in this study the Pune-Bremen DEWATS project scores lowest for state institutionalisation. It was largely planned, coordinated and implemented by non-state actors and institutions. The involvement of state institutions was limited to one pilot plant run by the PMC, plus the project advisor whose position was financed by state funds (see table 4).

Indicators: State Institutionalisation	Evaluation: DEWATS		
	Pune	Bremen	Total
Formal Approval by Local State Institutions	+-	--	+--
Formal Approval by (Sub)National State Institutions ((Sub)National)	--	--	--
Public Human Resources	+-	+-	+-
Public Financial Resources	+-	--	+--
Public Coordinator	--	--	--
Commitment of Local State Leaders	--	+-	+--

Table 4: Partnership Project DEWATS: Findings from Index State Institutionalisation

Formal Approval by Local and (Sub)National State Institutions

In the DEWATS partnership project the formal links to local state institutions were limited in both Bremen and Pune. Despite the intention to involve local state actors and to institutionalize the Pune-Bremen partnership more formally through the partnership MoUs, the cooperation between the two cities remained primarily based and focused on non-state institutions and actors. The DEWATS project reflects this continuity as the majority of the pilot plants were executed by non-state institutions while the local state bodies played a minor role. The partnership initiative did not require any state approval from the side of Bremen. Also the local state body from Pune, the PMC, remained rather passive in the DEWATS initiative and only engaged in one pilot plant that the PMC built together with BORDA. The other five DEWATS plants were set up in cooperation with the local NGOs BORDA from Bremen, MAM from Pune and Matru Mandir from Devruk without any involvement of state institutions. For a systematic dissemination of DEWATS in Pune much closer involvement of the local state institutions would have been required. However the PMC did not add DEWATS to its sanitation portfolio and as a result, no further DEWATS plants were built in the city.

The formal involvement of state institutions from higher policy levels was even more limited in the DEWATS partnership project. The DEWATS plants that were constructed as part of the partnership initiative did not require approval by institutions such as the state or national governments in Germany or India.

Public Human and Financial Resources

The DEWATS project was also largely implemented with non-state human and financial resources. The five NGO-led DEWATS plants in Pune and Devrukh were almost exclusively planned, coordinated and implemented by non-state actors. The NGOs BORDA, MAM and Matru Mandir provided or hired the employees required for building the plants. Only the project advisor was employed by the German government's international labour mobility program CIM. The CIM supports the placement of technical experts and managers from Germany in public and private institutions in developing countries to facilitate international cooperation in sustainable development. The project advisor explains that in contrast to most state-funded development cooperation positions which offer wages and benefits based on the host country's salary system, the CIM programs only provides its employees with compensation in line with local wages and no additional expenses or benefits. As a consequence, the project advisor's financial capacity as a CIM employee was very limited and he had to rely on Mahajani, the head of the IOA 21, to bring him to meetings and site visits on Mahajani's motorbike. The

project advisor did not have any funds to hire experts, such as engineers or draftsmen, to support him (Interview with DEWATS project advisor, 08-10-2013).
The financial resources for the five NGO-led plants were provided in full by the German NGO AWO. The PMC-run DEWATS plant was the only plant that was largely built using state resources. It was funded by the PMC and jointly constructed by BORDA and PMC employees (Interview with DEWATS project advisor, 15-09-2014).

Public Coordinator and Commitment of Local State Leaders

The attempt to strengthen the leadership and commitment of local state actors from Pune to the Pune-Bremen partnership through the setting up the MoUs and the IOA 21 had in general rather limited impact on the DEWATS project. The local public authority, the PMC, assumed little coordinating responsibility in the DEWATS partnership project and only supported one DEWATS plant financially.
The DEWATS projects were thus primarily run and coordinated by non-state actors. The NGO BORDA served as the major project coordinator in all six demonstration DEWATS plants in Pune and Devrukh. For the five NGO-led DEWATS plants BORDA set up collaborations with representatives from the local NGOs MAM (at the two DEWATS plants at the Sane Guruji Hospital in Hadapsar and at the plant at the children's hostel in Narodi) and Matru Mandir (at the two plants at the hospital in Devrukh). Gujar, chairman of the MAM and head of the Ayurveda hospital in Hadapsar, in particular turned out to be a crucial and reliable partner for BORDA and he took over substantial coordination work at the three BORDA plants at the MAM facilities (Interview with DEWATS project advisor, 08-10-2013).
The PMC's commitment to the DEWATS initiative and more generally to the Pune-Bremen partnership remained weak. The MoUs turned out to be largely declarations of intentions and the PMC showed little initiative to proactively develop concrete project proposals. After the adoption of the first partnership MoU in 1998, the PMC initially supported the establishment of the IOA 21 to mainstream and coordinate Local Agenda 21 processes in Pune, Pimpri Chinchwad and Bremen. According to an interviewee from the partnership association Forum Städtesolidarität Bremen-Pune the limited commitment of the PMC to the partnership was one of the main reasons why the IOA 21 failed to live up to expectations. This interviewee explains that the PMC kept sending unqualified employees to run the office that lacked the competences and influence to initiate projects in Pune (Interview with Forum Städtesolidarität Bremen-Pune, 27-02-2013). One task of the IOA 21 was to host regular stakeholder meetings to foster engagement in sus-

tainable development in Pune and Pimpri. An NGO representative involved explains that after a few meetings the participation of PMC representatives and stakeholders became sluggish and only low-ranked representatives were sent which eventually led to the meeting series being terminated (Interview with NGO Arbutus, 25-09-2013). The then head of the partnership association AFG, Mahajani, took over responsibility for the IOA 21 on a largely voluntary basis.

When the German project advisor arrived in Pune in 2005, he was surprised to see that the IOA 21 was hardly staffed. He adds that during the time he worked in Pune, the IOA 21 had no real function and practically no links to the city administration (Interview with project advisor, 08-10-2013). Another interviewee confirms that the IO A21 office was not well taken care of and gave a bad impression to visitors (Interview with journalist from Pune, 18-09-2013).

One regularly cited reason for the lack of long-term commitment of the PMC to the partnership projects is the frequent changes to the political leadership in Pune, specifically the commissioners and mayors (Interview with NGO MAM, 19-09-2013; Interview with Forum Städtesolidarität Bremen-Pune and LAFEZ, 27-02-2013; Interview with project advisor, 08-10-2013). An interviewee explains that in Pune political leaders change too regularly and while at the beginning of the DEWATS project the commissioner had shown some interest in the dissemination of the technology in Pune, his successors did not have the political will to follow up on this (Interview with NGO MAM, 19-09-2013). The project advisor confirms that during the time he worked in Pune the local commissioner did not show any interest in the DEWATS projects and that it was very difficult for the project advisor to gain access to talk to him (Interview with DEWATS project advisor, 15-09-2014). The project advisor adds that even for the one plant that the PMC funded the city administration did not take full ownership. He had the impression that the PMC only built this plant to prove their commitment to the partners from Bremen as the PMC had initially voiced strong interest in DEWATS, but had shown little engagement thereafter (ibid.).

A member of the AFG highlights that while the partnership protagonists have struggled to get state actors from Pune involved, they have never had any problems with the city administration and political leaders from Bremen (Interview with AFG, 26-11-2012). The interviewee from Forum Städtesolidarität Bremen-Pune confirms that Bremen's senate and mayors have shown continued support for LAFEZ's development cooperation work in Pune and other cities in the Global South. However, local politicians from Bremen have not played a major role in the DEWATS partnership project (Interviews with Forum Städtesolidarität Bremen-Pune, 27-02-2013 and 01-10-2014).

5.1.4.3 Local Partnership Entrepreneurs

A third hypothesized condition for the successful realisation of transnational urban partnership projects is the existence of engaged and well-networked local individuals who show a personal commitment to the projects' progress. In this study such individuals have been termed 'partnership entrepreneurs' (PEs). The PEs have been identified in project documentation and through interviews with partnership protagonists.

The DEWATS partnership project was driven by three individuals. Gunther Hilliges, the head of the LAFEZ in the Bremen city administration, co-initiated the project together with Dr. Gujar, the chairman of the NGO MAM and the Sane Guruji Hospital in Pune where the first DEWATS demonstration plants were built. The third PE was Vijay Mahajani, Bremen's main contact person in Pune. Mahajani was involved in the Pune-Bremen partnership and the DEWATS project as head of the partnership association and the IOA 21.

To summarize, the case study findings indicate that the three PEs fully or partly meet all six criteria for a successful PE, with Hilliges scoring highly in all six indicators (see table 5).

Indicators: Local Partnership Entrepreneur	Evaluation: DEWATS		
	Pune	Bremen	Total
Belief in Feasibility and Benefits	++	++	++
Investment of Resources	++	++	++
Ability to Convince Stakeholders	+-	++	++-
Internal Policy Network	+-	++	++-
External Policy Network	+-	++	++-
Partner City Policy Network	+-	++	++-

Table 5: Partnership Project DEWATS: Findings from Index Partnership Entrepreneur

Belief in Feasibility and Benefits, Investment of Resources and Ability to Convince Stakeholders

A precondition for engaging in a partnership project is the PE's belief in the feasibility and benefits of the project. This condition is met by all three PEs in the DEWATS project. Hilliges was the initiator and major driver of the Pune-Bremen city partnership. A representative from Terre des Hommes in Pune who knows Hilliges from several joint projects and exchange visits underlines Hilliges' critical role in the partnership. He states that without Hilliges and his personal engagement the partnership would never have come into existence nor been sustained for so long and that many of the projects depended heavily on Hilliges and his passion for realising partnership initiatives in Pune (Interview with NGO Terre des Hommes, 24-09-2013). Hilliges strongly supported the idea to set up DEWATS projects in Pune. The interviewee has observed several DEWATS plants in operation, including several in China, and he praises the ability of the technology to purify even pig farm effluents and provide clean water as an end product. Hilliges therefore considers the DEWATS technology as a suitable solution to reduce stress on urban systems caused by urbanization, particularly for newly built residential areas that have not been connected to a central sewage treatment system (Interview with Forum Städtesolidarität Bremen-Pune, 27-02-2013).

The two partnership entrepreneurs from Pune, Gujar and Mahajani, were also convinced that the DEWATS technology could help improve the sanitation situation in Pune. A member of the partnership association, the Pune-Bremen City Solidarity Forum, points out that the demonstration plants worked out so well because Gujar took great personal interest in the project, taking leadership and inviting BORDA to set up the first plants in the grounds of his hospital (Interview with NGO Arbutus, 25-09-2013).

Mahajani had been the main contact person for Hilliges and the LAFEZ long before the DEWATS project was initiated. Like Hilliges, Mahajani was personally engaged in fostering the Pune-Bremen partnership and during the project advisor's stay in Pune Mahajani helped him to establish local contacts for the DEWATS projects (Interview with project advisor, 08-10-2013). A representative of the PMC remembers that Mahajani advertised the DEWATS project at the PMC and states that since Mahajani passed away in 2010 the DEWATS initiative has lost impetus (Interview with PMC, 24-09-2013).

Another indicator measuring the commitment of PEs is their investment of private resources such as time, personal reputation or financial contributions in partnership activities. This indicator was also fulfilled by all three PEs of the DEWATS project. Hilliges for example donated 50,000 Indian Rupees which he collected from friends and colleagues at his birthday party in Mai 2004 for environmental projects in Pune (Forum Städtesolidarität Bremen-Pune e.V., 2006, 22). Since he retired and left the LAFEZ in 2005, Hilliges has also travelled several times to Pune at his own expense, as head of the partnership association Forum Städtesolidarität Bremen-Pune (Interview with Forum Städtesolidarität Bremen-Pune, 27-02-2013). An interviewee, who is more sceptical of Hilliges' approach to developing the city partnership in other regards, acknowledges that Hilliges' dedication to the Pune-Bremen partnership went beyond his professional engagement (Interview with local journalist from Pune, 18-09-2013).

The project advisor highlights that Mahajani was one of the few actors from the Pune side who permanently engaged in developing the partnership with Bremen both professionally and personally. He adds that Mahajani continued to work largely voluntarily at the IOA 21, even when his health deteriorated (Interview with project advisor, 08-10-2013). A representative of the PMC confirms that Mahajani's partnership engagement went far beyond his role as the head of the IOA 21 (Interview with PMC, 24-09-2013).

Likewise Gujar was personally committed to the successful construction of the DEWATS plants. The project advisor recalls that Gujar showed great dedication to the project and took personal responsibility for it even outside of working hours (Interview with project advisor, 08-10-2013).

As for the third indicator of successful PEs, the ability to "sell" the partnership project to different audiences, all three PEs are described as charismatic and respected personalities. According to interview partners the PEs have been crucial to the promotion of the DEWATS projects in Pune. A representative of the NGO MAM in Pune highlights Hilliges' skills to put forward ideas for joint partnership projects and his ability to bring different stakeholders together to implement projects (Interview with NGO MAM, 19-09-2013). Another interviewee confirms that Hilliges was a convincing advocate of partnership projects and able to drive projects forward in Pune, adding that this was related to the fact that Hilliges was an official representative of the Bremen government which provided him with an official status and a budget to fund partnership projects in Pune (Interview with journalist from Pune, 18-09-2013).

Gujar and Mahajani both earned respect in Pune due to their social engagement and their seniority. An interviewee highlights Gujar's good reputation in Pune thanks to his civil engagement (Interview with NGO Arbutus, 25-09-2013). The project advisor adds that Gujar had a very positive aura and a likeable personality (Interview with DEWATS project advisor, 08-10-2013). Also the PMC representative explains that Mahajani enjoyed trust and respect in the Pune city administration, which is crucial to being heard by city officials. This interviewee further describes Mahajani as a "genuine person" who spoke in a friendly way to everybody regardless of their rank and did not refer to his status as the IOA 21 representative (Interview with PMC, 24-09-2013).

The project advisor confirms that Mahajani was accepted and respected because of his age, but he also had the impression that the city administration did not always take him seriously, which became particularly apparent in the tramway project (Interview with project advisor, 08-10-2013; see more details in section 5.2.3.3).

Internal, External and Partner City Network

Another key competence of partnership entrepreneurs is their ability to access policy networks in their own city (internal policy networks), beyond the boundaries of their home city (external policy networks) and in the respective partner city (partner city policy networks). With regard to internal policy networks Hilliges could draw on close relations to local decision makers. He had good access to high ranking officials in the Bremen Senate and the Bremen city administration due to his position as the head of the LAFEZ. The fact that the LAFEZ was established as an official department in the administration enabled Hilliges to directly contact key political actors such as the mayor, the senator for environment or officials at the Department for Education (Interview with Forum Städtesolidarität Bremen-Pune, 27-02-2013). Hilliges has also developed good relations with local NGOs from Bremen that are active in the area of development cooperation. As a private member of BORDA, Hilliges was instrumental in bringing BORDA and MAM together in the DEWATS project (Interview with BORDA, 11-09-2014).

Mahajani's internal policy network was less comprehensive; he had close connections to local key political and industrial actors, but he lacked extensive relations with local NGOs (Interview with NGO Terre des Hommes, 24-09-2013). According to the PMC representative, Mahajani was well known in the city administration and in Pune's political circles, which was a prerequisite for raising awareness of partnership projects and speeding up processes. The PMC employee adds that Mahajani even had private contacts with the city's commissioners. According to this interviewee, well-networked individuals such as Mahajani are crucial to driving forward international partnership projects, as decision makers in Pune would not enter into discussions with someone from Germany without a mediator, particularly if financial issues were involved (Interview with PMC, 24-09-2013). A member of Bremen's partnership association confirms that the fact that Mahajani had close personal relations to local decision makers facilitated Bremen's partnership actors' access to the PMC. He explains that Mahajani was a member of one of the highest Indian castes which helped him to get heard by political leaders, while Mahajani's successors as heads of Pune's partnership association have found it more difficult to reach local politicians in Pune (Interview with Forum Städtesolidarität Bremen-Pune, 27-02-2013).

The interviewee from LAFEZ confirms that Bremen's contact to the PMC depended in great part on Mahajani's mediation, pointing out that when Mahajani passed away in 2010, contact to the PMC was lost (Interview with LAFEZ, 27-02-2013).

Mahajani's internal policy network was however more limited with regard to local NGOs in Pune, as the interviewee from Terre des Hommes points out (Interview with NGO Terre des Hommes, 24-09-2013). This interviewee adds that in Pune the divide between industry and NGO representatives is particularly strong which makes it difficult to bring both groups together in an initiative such as a city partnership (Interview with NGO Terre des Hommes, 24-09-2013). Another interviewee who attempted to widen the Pune-Bremen cooperation with a 2009 initiative to integrate more local NGOs found it difficult to convince Mahajani of this idea. This interviewee had at times even the impression that Mahajani tried to prevent the involvement of additional NGOs from Pune in the partnership (Interview with journalist from Pune, 18-09-2013).

Gujar on the other hand was well known among local NGOs such as Arbutus (Interview with NGO Arbutus, 25-09-2013) and Terre des Hommes (Interview with NGO Terre des Hommes, 24-09-2013). He was however generally less networked within political circles in Pune. The project advisor explains that Gujar primarily focused on NGO project work in his hospital and did not engage much in local politics (Interview with project advisor, 15-09-2014).

As for the external policy network, Hilliges has established an extensive network of contacts with NGOs and city representatives outside of Bremen that are active in development cooperation work. As part of his work as former director of Terre des Hommes Germany he was one of the pioneers of city-level North-South cooperation in Germany. During this time he also closely collaborated with partner NGOs from the Netherlands, with whom he founded the initiative "Towns and Development" to foster municipal development cooperation initiatives in 1986 (Interview with Forum Städtesolidarität Bremen-Pune, 27-02-2013). Hilliges has continued his civil engagement in national and international NGOs during his work at the LAFEZ and after his retirement. Today he serves as curator of the Stiftung Zukunftsfähigkeit (Foundation for Sustainability) and as board member of NGO Germanwatch which are both active in strengthening North-South relations in sustainable development.

As the head of the NGO MAM, Gujar also had frequent access to national and international NGOs with whom he collaborated on projects to promote sustainable development in rural India. The DEWATS partnership initiative strongly benefited from the long-term cooperation between the NGOs MAM and AWO in the five pilot DEWATS plants in Hadapsar, Narodi and Devrukh (Interview with NGO MAM, 19-09-2013).

Mahajani's external policy network was comparatively small. The project advisor states that in the case of the DEWATS project, Mahajani's contacts outside of Pune were rather limited (Interview with DEWATS project advisor, 08-10-2013).

With regard to the PEs' relations in the respective partner city Hilliges has travelled on numerous occasions to Pune; more than 50 times since 1976 (Forum Städtesolidarität Bremen-Pune e.V., 2006). He has established a network of close relationships with a group of citizens and NGOs from Pune, amongst others Mahajani (AFG), Gujar (MAM), and the NGOs Arbutus and Terre des Hommes. On his visits Hilliges also regularly met with representatives from the Chamber of Commerce as well as educational and health institutions in Pune and put them in touch with their counterparts in Bremen. While Mahajani facilitated contact with the PMC and Hilliges had an official position as a Bremen state representative Hilliges also enjoyed frequent access to Pune's political circles. A PMC employee emphasises that Hilliges was well-known in the PMC (Interview with PMC, 24-09-2013). Several interviewees however mention that the city partnership has remained rather elitist and that Hilliges has not managed to involve more influential NGOs and local citizens in partnership activities in Pune (Interviews with project advisor, 15-09-2014 and journalist from Pune, 18-09-2013).

Hilliges also served as the main contact person in Bremen for Mahajani and Gujar who had fewer direct relations to Bremen. Gujar had already collaborated with Hilliges and BORDA in the biogas projects launched at the beginning of the partnership in the 1970s and 1980s. According to the interviewee from MAM, Gujar visited Germany three times due to his connections with the AWO and visited Hilliges in Bremen at least once. This interviewee adds that they also met frequently in Pune, whenever Hilliges visited the city. On one of these exchange visits to Pune Gujar and Hilliges developed the plan to set up the DEWATS plants at the MAM facilities (Interview with NGO MAM, 19-09-2013). As the head of the partnership association and the IOA 21, Mahajani also met Hilliges regularly in Pune and he visited Bremen on many occasions. Other than Hilliges in Pune, Mahajani did not establish an extensive independent personal network of contacts in Bremen (Interview with project advisor, 15-09-2014).

5.1.4.4 Partnership Social Capital

The fourth assumed success factor for transnational urban cooperation is the development of social capital among the involved partnership protagonists and institutions from both partner cities.

According to the results of the index system the DEWATS project scores highest of the four analysed cases in partnership social capital. The collaboration stands out with regard to its extensive prior partnership activities and the high level of trust between the project partners from Germany and India, plus the protagonists' comprehensive experiences in intercultural exchange. It however scores lower with regard to stakeholder involvement and inclusion (see table 6).

Indicators: Partnership Social Capital	Evaluation: DEWATS		
	Pune	Bremen	Total
Prior Collective Action	/	/	++
Trust	++	+-	++-
(Perceived) Equality	+-	+-	+-
Intercultural Experiences	++	++	++
Intercultural Communication	/	/	+-
Involvement of Leadership Networks	+-	--	+--
Inclusion & Citizen Participation	+-	--	+--

Table 6: Partnership Project DEWATS: Findings from Index Partnership Social Capital

Prior Collective Action, Trust and (Perceived) Equality

The actors involved in the DEWATS project could draw on comprehensive experiences from a multiplicity of prior environmental and other partnership projects conducted under the head of the Pune-Bremen city cooperation. Particularly, decentralized waste management has been one of the partnership's focus areas since its beginnings in the 1970s. Several of the actors (Hilliges and Gujar) and NGOs

(BORDA and MAM) collaborating in the DEWATS project had already been involved in the development of small-scale biogas plants in rural Pune in the late 1970s and early 1980s. In addition, over the more than 25 years of partnership activities prior to the DEWATS project, the two cities have jointly set up a wide array of small and medium-sized projects, dozens of conferences and workshops; and hundreds of citizens from Pune and Bremen have participated in mutual exchange visits (see more detailed information about the partnership history in section 5.1.1). As a result of this extensive prior collective action and personal exposure to both cities several key actors in the DEWATS projects (amongst others the three PEs, Hilliges, Gujar and Mahajani) were already familiar with the cultural and other context differences between Pune and Bremen, which facilitated the exchange.

The prior partnership experiences also helped build a generally trusting and cooperative atmosphere among the protagonists who were engaged in the partnership from Bremen and Pune. Many members of the partnership associations and local citizens have developed personal relationships. The head of the partnership association AFG calls them "small bridges of friendship" (Interview with AFG, 26-11-2012). A concrete example of the trusting relations in the Pune-Bremen partnership is the fact that the LAFEZ has delegated the distribution and monitoring of its annual project funding for 12-14 NGOs from Pune to the partnership organisation AFG from Pune (Forum Städtesolidarität Bremen-Pune e.V., 2006). The interviewees from LAFEZ and Forum Städtesolidarität Bremen-Pune highlight that the AFG has always been reliable and that Bremen has never regretted this decision (Interview with Forum Städtesolidarität Bremen-Pune and LAFEZ 27-02-2013).

The interviewee from the partnership association AFG emphasises that the partnership actors from Pune consider Hilliges a close friend and that they offer him a high degree of respect and gratitude due to his continued commitment to the city of Pune (Interview with AFG, 26-11-2012)

The interviewee from the NGO Arbutus from Pune points out that personal relations between partnership protagonists are an important success condition for a transnational urban cooperation such as the Pune-Bremen partnership (Interview with NGO Arbutus, 25-09-2013). The project advisor confirms that according to his experiences many partnership ideas and projects were developed as a result of personal exchange between Hilliges and his contacts in Pune (Interview with DEWATS project advisor, 08-10-2013).

While the non-state actors in the partnership have established trusting relationships, the confidence of most protagonists (from both Bremen and Pune) in Pune's political decision makers and city administration has been more limited.

An interviewee gives the example of the PMC promising to support NGO partnership activities with the same amount that the LAFEZ donates annually. This promise was however not kept (Interview with AFG, 26-11-2012).

The DEWATS project reflected the overall partnership with regard to widely trusting relations between non-state actors, whereas the confidence in the competences and reliability of Pune's state representatives remained more limited. The project advisor recalls that the cooperation between BORDA, Gujar and himself benefited from a high degree of mutual trust (Interview with DEWATS project advisor, 08-10-2013).

The collaboration between the project advisor and the PMC ran less smoothly, which the project advisor explains was in part due to the lack of competence and know-how of the employees whom the PMC sent to explore new sites for potential DEWATS plants. He states that these PMC employees often suggested sites to him that were unsuitable for DEWATS plants and he had the impression that they had not fully understood how the DEWATS plants worked (ibid.).

A key challenge that many partnerships between Northern and Southern cities face is establishing equality between the protagonists. Often differences in levels of wealth, education and access to information are substantial and they can lead to one-sided exchange or at least to the perception of it. The interviewee from Terre des Hommes in Pune, who has participated in several activities as part of the Pune-Bremen cooperation, explains that both his NGO and the Pune-Bremen partnership have made great efforts to ensure equality in their North-South relations but that this has not always been successful. According to this interviewee, the major barrier to equal relationships is the one-sided distribution of funds. These funds are usually provided by the Northern partner, leading to power imbalances in the partnerships. These imbalances are often exacerbated by the fact that in the recpient city or organisation different actors are competing for the funds. In the case of the Pune-Bremen cooperation the interviewee explains that the partnership entrepreneurs Hilliges and Mahajani achieved parity in their exchange but that many other actors from Pune were primarily interested in getting their projects funded, which prevented more equal cooperation. The interviewee states that one-sided funding can in particular disturb partnerships such as the one between Pune and Bremen that aim to collaborate on an equal footing (Interview with NGO Terre des Hommes, 24-09-2013).

The LAFEZ and Bremen's partnership association, the Forum Städtesolidarität Bremen-Pune, are aware of this funding dilemma. One reason for delegating the distribution of the NGO funds to the AFG was to weaken Bremen's role as donator in the partnership (Interview with Forum Städtesolidarität Bremen-Pune and LAFEZ, 27-02-2013).

The establishment of the jointly financed IOA 21 office in Pune was another attempt to strengthen equality in the partnership and give more decision-making responsibility to the actors from Pune. In the 2003 MoU the partner cities agreed that the IOA 21 "[...] will be guided by decisions from a Board of Directors, which will represent important stakeholders from Pune and Pimpri Chinchwad". Furthermore, the Board should "[...] decide on the staff and will be accountable for the use of funds shared by the three cities." (Forum Städtesolidarität Bremen-Pune e.V., 2006, 13). However, after a promising start, the IOA 21 lost impetus and it eventually failed to live up to expectations, as it became neither a permanent contact point for Bremen nor an influential driver of partnership projects or Local Agenda 21 processes in Pune.

In the DEWATS project the funding dilemma was lessened by the fact that the project funds for the demonstration plants were not provided by the city of Bremen, but by the German NGO AWO and the PMC. As a result the project partners collaborated largely on an equal footing, as the project advisor points out (Interview with DEWATS project advisor, 08-10-2013).

Intercultural Competence and Communication

As elaborated in section 4.4 the two dimensions of intercultural experiences and communication have been added to the partnership social capital index due to their relevance for North-South cooperation.

At the start of the DEWATS partnership project all major protagonists (Hilliges and the LAFEZ, BORDA, the project advisor, Gujar, Mahajani and the PMC) had experience in international projects and exchange. India has been one of the LAFEZ's major partner countries for development cooperation since its foundation in 1979. The LAFEZ employees have conducted a multiplicity of environmental, poverty reduction and education projects in India, many in cooperation with BORDA (for an overview see Freie Hansestadt Bremen, 2005). Based on the experiences in international projects, the LAFEZ has specialized in intercultural communication and management and published multi-lingual manuals for intercultural trainings for which there was, according to the LAFEZ, a high demand from NGOs, academic institutions and private businesses (Interview with LAFEZ, 27-02-2013).

The biogas projects in rural Pune in the 1970s and 1980s were the first cases of cooperation between the LAFEZ, BORDA and Gujar and his NGO MAM. However Gujar also had a long working relationship with the AWO who funded several DEWATS demonstration plants. According to the project advisor, the DEWATS project benefited from the fact that Gujar and the MAM had collaborated with German and Bremen partners before (Interview with DEWATS project advisor, 08-10-2013). He himself had never worked in India or Asia before he moved to Pune but he could draw on his intercultural experiences from development cooperation work in Nicaragua and Belize where he had set up several decentralized biogas and waste management projects prior to his move to Pune.

On the PMC side Commissioner Sanjay Kumar and Mayor Dipti Chaudhari, who signed the 2003 partnership MoU and initiated the PMC-run DEWATS demonstration plant in Hadapsar, visited Bremen in 2004 (Forum Städtesolidarität Bremen-Pune e.V., 2006). Pune's political representatives had also been exposed to international exchange via Pune's other sister city agreements with Okayama (Japan) and San José (USA). The partnership entrepreneur Mahajani spent two years in Aachen in Germany as a student. When he returned to Pune he co-founded the AFG and became the key contact for Hilliges and the LAFEZ in the Pune-Bremen partnership (Interview with AFG, 26-11-2012). Mahajani also travelled several times to Bremen and other parts of Germany as part of the Pune-Bremen partnership as well as on his own initiative and he was therefore well aware of cultural differences between Germany and India (Interview with PMC, 24-09-2013).

With regards to partnership communication, the fact that the majority of partnership actors spoke the same language facilitated the exchange. Many partnership actors from Pune, among them Mahajani and his AFG fellows speak German as a result of their stays and contacts in Germany. Pune was the first Indian city to introduce the teaching of German to school children in 1914 and today Pune is India's number one city for German language education in India. The AFG has about 200 to 250 members in Pune who have been involved in the Pune-Bremen partnership. According to the interviewee from AFG, the fact that many partnership actors from Pune speak German has been of great benefit to the partnership (Interview with AFG, 26-11-2012). This interviewee adds that today most German partners speak English so that language is no barrier to the partnership (Interview with AFG, 26-11-2012). Another interviewee agrees with this statement, emphasising that although language may have been a challenge at the beginning of the Pune-Bremen partnership, in recent years all the Germans involved could speak English reasonably well and language is no longer an issue (Interview with journalist from Pune, 18-09-2013).

Despite the high levels of common language skills in the partnership, some additional communication barriers had still to be overcome in the DEWATS project. The project advisor points out that sometimes linguistic and cultural misunderstandings led to the local engineer making mistakes in the project drawings. Even if the project advisor and his BORDA colleagues were able to correct them and they never resulted in any construction errors, the project advisor suggests that future cooperation should consider solutions to avoid such communication pro blems (Interview with DEWATS project advisor, 08-10-2013).

Involvement of Leadership Networks, Inclusion, and Citizen Participation

An important condition for strengthening social capital in transnational urban cooperation is the involvement of informal leadership networks and the inclusion of minorities and underprivileged groups, into partnership activities.

Since its beginnings in the 1970s, the Pune-Bremen city partnership has been driven by a rather stable group of engaged local citizens with Hilliges and Mahajani and the partnership associations from Pune and Bremen at the centre. A key objective and unique characteristic of the partnership has been the focus on civil society and private actor involvement. The partnership between the two cities was launched by NGO initiatives such as the Terre des Hommes projects for handicapped children, poverty reduction and public health and the biogas projects conducted by the NGOs BORDA from Bremen and MAM and UNDARP from Pune (see section 5.1.1). Throughout the initial 25 years of cooperation prior to the initiation of the DEWATS projects, Hilliges and his partnership supporters conducted several attempts to facilitate and strengthen the involvement of local NGOs, educational institutions and business entrepreneurs from Pune and Bremen. For example the NGO Arbutus from Pune has been engaged in activities since the early years of the partnership and set up a number of environmental and social education projects in collaboration with Hilliges and the LAFEZ from Bremen. Moreover, the LAFEZ and the AFG have established a long-term cooperation with selected partner NGOs from Pune who receive regular financial support. In addition to the NGO activities, Hilliges and the LAFEZ encouraged local universities and the chambers of commerce to sign partnership MoUs. All the same, the group of permanently active partnership protagonists remained relatively small. Cooperation with the chambers of commerce never really got off the ground and apart from an India roundtable that was established at the Bremen Chamber of Commerce not much business exchange has taken place as part of the partnership (Interview with Forum Städtesolidarität Bremen-Pune, 27-02-2013). Bremen and Pune's universities have collaborated more closely and set up joint exchange programs and regular summer schools. However, academic cooperation has remained in the hands of just a few academics and has seen great variations in its intensity, as one interviewee stated off the record. Also the number of local NGOs who have engaged in the partnership has been limited and several influential NGOs such as Parisar from Pune have not engaged in the city cooperation. Particularly in Pune the partnership has struggled to overcome existing divides between NGOs, industry and political decision makers. An interviewee explained that the fact that Mahajani and most AFG members were industry representatives has prevented a wider involvement of local NGOs from Pune (Interview with NGO Terre des Hommes, 24-09-2013). An interviewee from the PMC adds that the general atmosphere between NGOs and the city administration is likewise tense (Interview with PMC, 24-09-2013).

Two significant initiatives to widen the partnership network have been unsuccessful. The IOA 21 targeted broad stakeholder involvement in the development of a Local Agenda 21 strategy for Pune and Pimpri Chinchwad. But after showing initial interest, local business and political leaders' support for the initiative was fading and both the strategy and the IOA 21 lost impetus (Interview with NGO Arbutus, 25-09-2013). Also another attempt to extend the NGO cooperation between Pune and Bremen failed, this time due to a lack of response from stakeholders from Bremen, as an interviewee involved in this initiative points out (Interview with journalist from Pune, 18-09-2013).

The DEWATS project reflects the general partnership with regard to the limited direct involvement of local stakeholder networks. In the implementation of the demonstration plants the local NGOs BORDA, MAM and Matru Mandir took over responsibility for the execution. But apart from these NGOs, the DEWATS project had very few links to local stakeholder networks which proved to be a barrier for a wider dissemination of the technology in Pune. The project advisor highlights that close contact with other local NGOs would have been crucial for a wider project success, as they provide an insight into local politics and can facilitate access to the key decision makers (Interview with DEWATS project advisor, 08-10-2013). The only actor from Pune who was partly involved on an informal level in the DEWATS project was Professor Venkat Gunale from the Department of Botany at the Pune of University. Gunale took a personal interest in the DEWATS technology and joined several DEWATS events in Bremen and Pune (Interview with DEWATS project advisor, 08-10-2013). On the Bremen side no additional local actors took part in the DEWATS partnership project.

With regard to the indicator of inclusion in this partnership initiative, underprivileged communities from India were a major target group for the DEWATS project; all six demonstration plants were built at facilities that support marginalized groups. BORDA built three plants in collaboration with the NGO MAM from Pune which provides medical treatment and vocational training for tribal communities from rural Pune. Two additional plants were set up at an orphanage in Devrukh that is run by the local NGO Matru Mandir which also works with underprivileged rural communities, in particular the disadvantaged social groups of the Dalits and the indigenous Adivasi community (BORDA, 2005). Also the PMC-run plant was built as part of a new residential area for former slum dwellers in Hadapsar and was intended to serve as a model project for wider dissemination of DEWATS in urban poverty zones in India (Freie Hansestadt Bremen, 2005). The underprivileged groups targeted were however not actively involved in the project planning and implementation. From Bremen no underprivileged actors or communities were included in the DEWATS initiative.

5.2 Case Study Pune – Bremen II: Transferring the Bremen Tramway System to Pune

5.2.1 Partnership Project Tramway: Content and Process

The second major partnership project that was pursued as a follow-up to the Pune-Bremen 2003 partnership MoU was an initiative to transfer the Bremen tramway system to the Pune metropolitan area.

According to the secretary of the partnership organisation AFG, improving local transport conditions has become a crucial challenge in Pune. Over the last decades, the number of private cars and motorbikes has been steadily growing in the city due to urbanisation and industrialisation, which has put the road infrastructure under immense pressure and led to congestion and increased air pollution. The AFG representative concludes that Pune urgently needs to strengthen its public transportation in order to reduce the use of private vehicles (Interview with AFG, 26-11-2012). Vijay Mahajani, the then chairman of the AFG, and Gunther Hilliges, then head of the LAFEZ, had in 2003 already identified public transportation as a major field of cooperation for the Pune-Bremen partnership. They were particularly impressed by the idea of introducing a tramway system in Pune in order to strengthen public transport and foster sustainable and climate-friendly mobility.

Even though existing tramways are very rare in Indian cities (in fact there is only one rather run-down tramway system in Kolkata, built by the British colonial rulers at the beginning of the 20th century), Hilliges and Mahajani managed to pique the interest of local decision makers from Pune and its adjacent neighbouring city Pimpri Chinchwad in the project. An interviewee, who worked at the PMC's transport department at that time, confirms that it was the two partnership protagonists who introduced the tramway concept to the city administration in Pune (Interview with PMC, 24-09-2013). The idea was developed during the visit of a delegation from Pune to Bremen in April 2003 where a PMC city planner met with representatives of the Bremen Senator for Building, Environment and Transport, and the tramway company Bremen Straßenbahn AG (BSAG). The BSAG offered the Pune delegation its assistance in the construction of a tramway network connecting Pune and Pimpri Chinchwad and suggested concrete collaboration in the areas of wagon design, support with the construction and staff training (Freie Hansestadt Bremen, 2005). In June 2004 another Pune delegation, led by Pune's Mayor Deepti Chaudhari, visited Bremen to get insight into the workings of Bremen's tramway system. Hilliges and Georg Drechsler, head of the BSAG, took the partnership delegation on a test ride in the Bremen tramway.

The final decision to initiate the tramway transfer was taken during a visit by Hilliges and Stefan Boltz, the head of the Town Planning Department of the Bremen Senate, to Pune in December 2004. The two representatives from Bremen met with the commissioners from Pune and Pimpri Chinchwad who decided to send a local urban transport planner to Bremen to study the workings of the tramway. The urban transport planner had worked with the Pune and Pimpri city administrations before and was also interested in establishing new contacts with Bremen and Germany for his own business. In March 2005, the planner went to Bremen to join a workshop where he was introduced to the tramway technology and its operation and maintenance by Boltz and Jörg Monsees, managing director of Consult Team Bremen (CTB), a subsidiary of the BSAG. After the workshop Monsees and Boltz visited Pune to assess the local context together with Friedrich Steiger, the executive director of BGS Ingenieurconsult International in Frankfurt who was appointed by CTB to join the project as an additional consultant (Interview with urban transport planner, 11-10-2013). The team concluded that a tramway could be principally built in Pune and Pimpri Chinchwad, but in order to test the technical feasibility a more detailed analysis was required. They submitted a financial proposal for a Detailed Project Report (DPR) which was accepted by the commissioners of Pune and Pimpri Chinchwad and financially approved by the municipal bodies' Standing Committees. The PMC and PCMC agreed to share the costs of 50,000 Euro budgeted for preparing the DPR (Interviews with PMC, 24-09-2013, urban transport planner, 11-10-2013 and Forum Städtesolidarität Bremen-Pune, 01-10-2014).

In July 2005 a delegation of 45 officials from Pune and Pimpri Chinchwad, led by Pune's Mayor Chaudhari and the commissioners from both cities, visited Bremen on July 6 the Memorandum of Understanding for the tramway collaboration was signed as part of a senate reception in the Bremen town hall (Forum Städtesolidarität Bremen-Pune e.V., 2006). Over the following two years, a team including the German consultants Monsees and Steiger, Boltz from the Bremen city administration and the urban transport planner from Pimpri Chinchwad jointly set up the DPR. The team collected the required data for the report on several exchange visits to Bremen and Pune. They conducted a total of eight to ten site visits to both cities of four to five days each (Interview with urban transport planner, 11-10-2013). The urban transport planner was mainly responsible for coordinating the project from the side of Pune and Pimpri Chinchwad and he was also in charge of developing the proposed tramway alignment map. Boltz, Monsees and Steiger took over responsibility for the technology, town planning as well as financial aspects in the DPR.

In July 2007 the team submitted the completed DPR to the municipal corporations of Pune and Pimpri Chinchwad and presented its core findings to local decision makers and the press media. The DPR document comprises a total of 260 pages (Consult Team Bremen, BGS Ingenieurconsult International, 2007). It analyses the financial, socio-economic and environmental conditions for a tramway network in Pune and proposes concrete technical options, planning steps and favourable sites. The urban transport planner points out that from the German perspective the document would still not fulfil all the requirements of a DPR as a "detailed blueprint to start the work" (Interview with urban transport planner, 11-10-2013). He explains that in India such blueprints are not common and rather he would define the report as a 'feasibility study' according to German standards. In addition to the study, the urban transport planner submitted a detailed alignment map. The map includes the proposed tramway routes connecting all major areas in Pune and Pimpri Chinchwad and contained several underground sections in the inner city of Pune. The six lines were to be erected in two phases (see figure 1).

The Bremen tramway company BSAG showed a strong interest in the project, offering Pune several older tramway models for an inexpensive price. The partners from Bremen had even already arranged the transportation of these models to Pune with local shipping companies from Bremen (Interview with Forum Städtesolidarität Bremen-Pune, 01-10-2014).

Despite this extensive preparatory work, the Pune tramway system has never been realised. The municipal corporations of Pune and Pimpri Chinchwad forwarded the DPR to the Maharashtra state government for approval as the Indian cities neither had the legal authority nor the financial capacity to build a transport infrastructure project of such a scale alone. The state government however rejected the proposal. The main reason for the refusal was the fact that many Indian cities such as Delhi, Kolkata and Mumbai were fostering metro systems at that time and the Maharashtra state government was proposing to set up a metro in Pune. After the state government's rejection of the proposal, local political support also waned. The former mayor of Pune and commissioners of Pune and Pimpri Chinchwad who initially supported the tramway in 2003 had been replaced and their successors did not fight for the project.

116　　　　　　　　　　　　　　　　　　　　　　　　　　5 Within Case Analysis

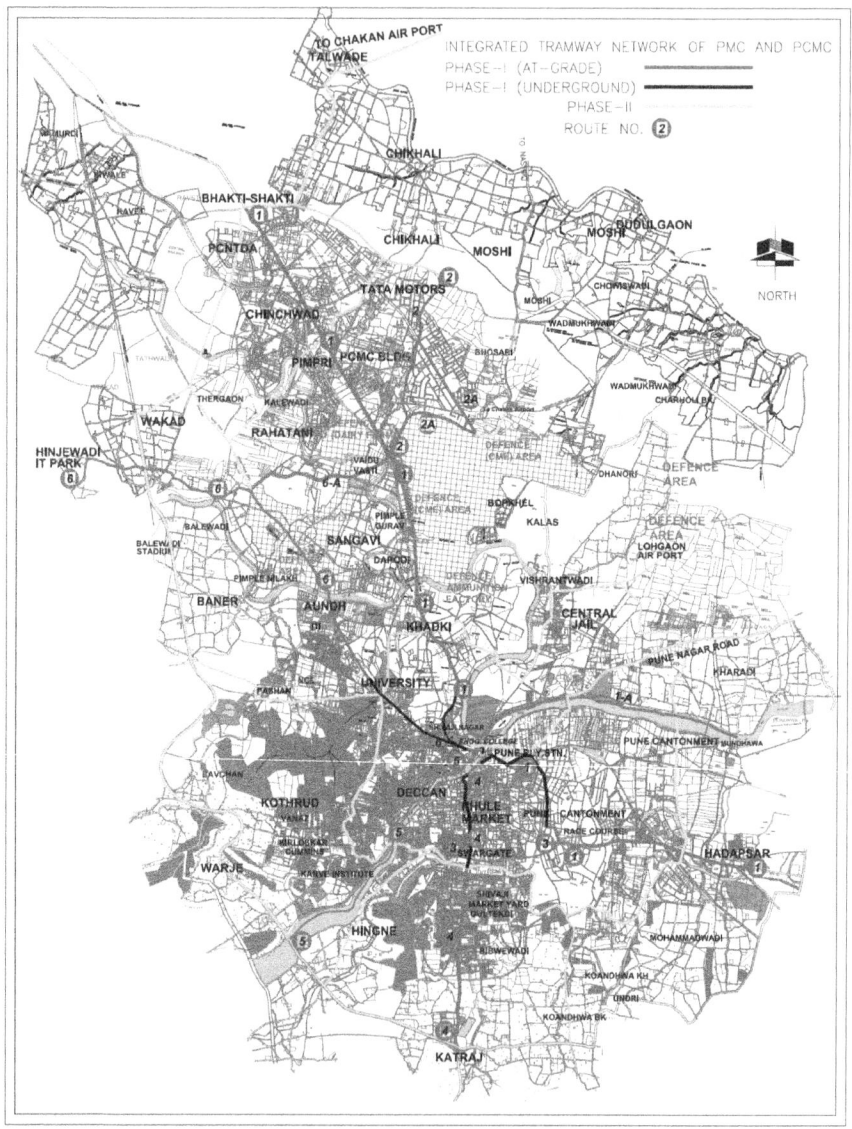

Figure 1:　　Tramway alignment plan for Pune and Pimpri Chinchwad; source: Sakhalkar (2007)

The PMC eventually even refused to pay its share of 25,000 Euro for the preparation of the DPR to the German consultants. When Monsees left the CTB and Vijay Mahajani, the leading advocate of the tramway transfer in Pune, passed away in 2010, the project also lost two major partnership proponents. CTB, Hilliges and the head of the NGO Arbutus (who succeeded Mahajani as Bremen's main partnership contact from Pune) made several fruitless attempts to convince the corporations to transfer the amount and revive the project, but they remained unsuccessful (Interviews with NGO Arbutus, 25-09-2013, urban transport planner, 11-10-2013 and Forum Städtesolidarität Bremen-Pune, 01-10-2014).

5.2.2 Partnership Project Tramway: Outcomes

Evaluation of the project's outcomes shows that the tramway project largely failed to realise its targets (see table 7). It only performs well in perceived mutuality in the partnership and it had some partial success in local capacity building as well as post-project sustainability.

Success Indicators	Project Evaluation: Tramway		
	Pune	Bremen	Total
Implementation Achievements	/	/	--
Budget and Schedule Performance	/	/	--
Local Capacity Building	+-	+-	+-
Benefits for Target Group	/	/	--
Impact	--	--	--
Mutuality	/	/	++
Post-project Sustainability	/	/	+-

Table 7: Partnership Project Tramway: Findings from success index Implementation Achievements

The only tangible outputs of the tramway project were the DPR and the tramway alignment map which the team made up of the urban transport planner, Monsees and Steiger submitted to the Pune and Pimpri Chinchwad city corporations in July 2007. These two documents were promising first steps in the attempt to transfer the Bremen tramway system to Pune and Pimpri. However, the practical implementation never progressed and the project ended shortly afterwards due to the lack of support from both the state and local governments. The interviewee of the PMC confirms that the project virtually stopped with the completion of the DPR (Interview with PMC, 24-09-2013). While the urban transport planner is still convinced that a tramway would have a future in a multi-modal transport system in Pune (Interview with urban transport planner, 11-10-2013), the interviewed PMC representative is more sceptical due to competition from other modes of transport (Interview with PMC, 24-09-2013). In fact, the Pune and Pimpri Chinchwad city administrations have decided to respond to the ever-increasing traffic by investing in recent years in street flyovers, setting up a Bus Rapid Transit (BRT) system and advancing a city metro. However, the BRT system has faced many problems and is not running properly and the metro has still not been built. As a result, public transport in the two cities is still in a poor condition and Pune is continuously ranked "among the most polluted [cities] in the country with rising noise, air and visual pollution" (Jadhav, Times of India, 7-10-2013).

Budget and Schedule Performance

The tramway partnership project faced major problems with regard to its budget. Initially, the project budget for the preparation of the DPR was fully approved by the then commissioners and Standing Committees of Pune and Pimpri Chinchwad. The city administrations engaged Monsees, Steiger and the urban transport planner to set up the DPR and the PCMC transferred its initial share of 25,000 Euro to the German consultants at the beginning of their engagement in 2005 (the urban transport planner contributed to the project voluntarily and was not paid for his work). However, after the DPR was rejected by the Maharashtra state government in 2007, the PMC refused to pay their share of 25,000 Euros because they no longer saw any benefits in conducting the study (Interview with PMC, 24-09-2013). As the German consultants had completed their work and delivered the study which the city administrations had commissioned them to carry out, neither they nor Hilliges could understand that the remaining amount was not paid. Hilliges and the CTB repeatedly contacted the city administrations to remind them of the outstanding payment, but the sum was never transferred.

As for the time schedule performance of the project, the urban transport planner states that the DPR was handed in on time. However, as the Maharashtra state government refused to approve the DPR, no further concrete planning was conducted to implement the project (urban transport planner by email, 09-09-2014).
Local Capacity Building

Local capacity building is one of the few areas in which the tramway projects delivered some achievements. Through the process of discussing the idea to introduce a tramway and preparing the DPR study, several actors improved their knowledge of tramway technology and how it could work in Pune and Pimpri Chinchwad. Boltz, Monsees and Steiger were already experts in the field of tramway infrastructure and city planning, but it was a new experience for them to adapt the technology to the context of urban India as they had never worked in the country before. Also their local project partner, the urban transport planner from Pimpri Chinchwad, visited Bremen and other German cities for the first time as part of this initiative where he was introduced to the tramway systems. Today, the urban transport planner is an advocate for tramway systems and he still hopes that Pune and also other Indian cities will adopt the technology in the coming years (Interview with urban transport planner, 11-10-2013). Also other actors from Pune who have been involved in the Pune-Bremen partnership became aware of tramway technology through the project and developed detailed know-how about tramway technology as well as an understanding of whether it fits into Pune's local context or not (Interviews with PMC, 24-09-2013, NGO Arbutus, 25-09-2013 and NGO Terre des Hommes, 24-09-2013,).

Apart from this individual learning which has been highlighted as an important but often undervalued outcome of transnational urban development cooperation projects (Devers-Kanoglu, 2009), the tramway partnership project also fostered institutional reform in the merging of the transport departments of the Pune and Pimpri Chinchwad city administrations. The DPR points out that the merger was conducted to ensure better coordination of the tramway network and the then separately managed bus systems of both cities (Consult Team Bremen, BGS Ingenieurconsult International, 2007, 140). Despite the failure of the tramway transfer this merger was still executed and today the Pune and Pimpri Chinchwad city administrations run a joint bus system which connects both cities (Interview with NGO Terre des Hommes, 24-09-2013). The urban transport planner notes that also the project team's research for the tramway alignment map was not completely redundant as the draft plan for the new metro in Pune has been oriented on the routes that he and the German consultants had recommended for the tramway (Interview with urban transport planner, 11-10-2013).

Benefits for Target Groups

One core aim of the tramway project was to connect industrial areas located outside of the cities to the public transport network. The idea was to provide workers living in the city centres with an economical, safe and environmentally friendly way of accessing their workplaces and thereby replacing private cars and motorbikes which are congesting the main arterial streets. The DPR highlights these benefits for the local workforce and the city population which is suffering from ever increasing traffic problems. The study promotes the tramway as a means to strengthen efforts by the city administrations to decentralise the city development and provide better access to the suburban industrial areas (Consult Team Bremen, BGS Ingenieurconsult International, 2007, 140).

However, as the tramway project was not implemented, the target groups, i.e. local workers and other members of the local population were not able to enjoy these benefits.

Positive Impact beyond the Partnership Project

For the German consultants, Monsees and Steiger, part of the motivation to join the partnership project and build a tramway in Pune was to demonstrate the feasibility of running modern tramways on the Indian subcontinent and thereby foster the diffusion of the technology to other Indian cities (Interview with urban transport planner, 11-10-2013). But ten years after the project was initiated, neither Pune nor any other Indian city has included modern tramways into their public transport portfolio and to this day Kolkata remains the only Indian city running a tramway system. More generally, none of the interview partners questioned as part of this research project was aware of any wider impact of the tramway project in or beyond the Pune metropolitan area, apart from the capacity building outlined above (see above).

(Perceived) Partnership Mutuality

Partnership mutuality is the only success indicator where the tramway project scores relatively highly. In contrast to many prior partnership projects that had been funded by institutions from Bremen or Germany (the DEWATS project for example), in the case of the tramway project the Pune and Pimpri Chinchwad municipal corporations agreed to share the costs of setting up the DPR. The interviewed journalist from Pune who observed the partnership initiative explains that the project even had the potential to bring more equality and mutual benefits to the entire Pune-Bremen partnership by mobilizing local business interests. He explains that the tramway project could have been an important step to overcome the one-sided donor-recipient relationship as it could have proven to the stakeholders from Bremen that they have economic benefits from the partnership with Pune (Interview with journalist from Pune, 18-09-2013).

All three consultants involved in the project were hoping to initiate business relationships through the partnership and in fact interviewed urban transport planner and Steiger became business partners in other projects as a result of their collaboration in the tramway DPR (Interview with urban transport planner, 11-10-2013). After a promising start to an equal partnership the abrupt ending to the tramway project led to much disappointment in both Bremen and Pune.

Post-project Sustainability of the City Partnership

The failed implementation of the tramway project led to a lot of frustration, particularly among the actors from Bremen. The German project partners eventually stopped pursuing the project and the remaining funding. The interviewees from the Forum Städtesolidarität Bremen-Pune and LAFEZ point out that today most partnership actors from Bremen prefer not to talk about the project anymore as they find it too painful and frustrating (Interview with Forum Städtesolidarität Bremen-Pune and LAFEZ, 27-02-2013). Also an interviewee from Pune refers to the tramway initiative as a "disturbing example of city-to-city cooperation" (Interview with NGO Arbutus, 09-12-2012). This interviewee points out that the failed tramway transfer had an adverse effect on the wider partnership between the two cities. He recalls that the project got a lot of negative press in Pune and that for a long time local journalists chose to focus on the failed tramway initiative when reporting on the partnership (ibid.).

As elaborated in the DEWATS case study, political actors from Pune largely withdrew from the Pune-Bremen partnership after the end of the DEWATS and tramway projects (see section 5.1.3). The LAFEZ made several attempts to contact the Pune city administration to pave the way for a new partnership MoU, but it received no response. Finally, the LAFEZ gave up as it did not have the capacity to keep the exchange with Pune's political institutions alive through regular exchange visits and it has since limited their engagement to NGO projects (Interview with LAFEZ, 27-02-2013).

5.2.3 Partnership Project Tramway: Explanatory Factors

5.2.3.1 Knowledge Exchange Strategy

As for the index scoring of the first explanatory variable, the knowledge exchange strategy, the tramway project performed well with regard to prior program evaluation and in the addressing of intrinsic interests in the partner cities. Similar to the DEWATS project, the tramway initiative scores low in the indicator measuring external moderation and facilitation as it was conducted without outside help. It partly fulfils the remaining three indicators of prior context evaluation, incremental and systematic transfer, continual interaction and the utilisation of policy windows (see table 8).

Indicator: Knowledge Exchange Strategy	Project Evaluation: Tramway		
	Pune	Bremen	Total
Prior Program Evaluation	/	/	++
Prior Context Evaluation	/	/	+-
Systematic Cooperation	/	/	+-
Continuity of Interaction	/	/	+-
External Moderator/Facilitator	/	/	--
Policy Window	/	/	+-
Intrinsic Interests	+-	++	++-

Table 8: Partnership Project Tramway: Findings from Index Knowledge Exchange Strategy

Prior Program and Context Evaluation

The tramway project particularly excelled with regard to the extensive prior evaluation of the tramway technology. The urban transport planner from Pimpri Chinchwad who had not worked with tramways before, was introduced to the technology on a number of site visits to Bremen. He recalls that during his first visit to Bremen he joined a workshop on the technical set up of the Bremen tramway system, organised by Stefan Boltz from the Bremen city administration (Interview with urban transport planner, 11-10-2013). Monsees, managing director of the traffic planning consultancy, the CTB, and Steiger, executive director of the engineering consultancy, the BGS Ingenieurconsult International, brought their extensive experience in international infrastructure planning and tramway construction to the project. They took over the main responsibility for the technical side of the project and they dedicated a full chapter of 30 pages in the DPR to explain the tramway technology to the project partners. Chapter 5 in the document provides detailed information about tramway car components and it explains tramway infrastructure elements such as energy supply, tram stops and storm water drainage. There is a particular focus on the different options to incorporate tramway lanes into the road network; these options range from integrating tramway lanes into existing roads to constructing new lanes fully or partly separated from the roads (Consult Team Bremen, BGS Ingenieurconsult International, 2007, 87-117). Multiple pictures and cross section graphics illustrate the distinct approaches to construct tramway lanes.

The DPR also provides comprehensive information about the specific local context conditions that determine the feasibility of transferring a German tramway system to Pune and Pimpri Chinchwad. According to the urban transport planner, a key accomplishment of the DPR preparation was that the project team conducted Pune's first ever origin-destination (OD) survey, measuring the travel distances and times of Pune's transport system (Interview with urban transport planner, 11-10-2013). The OD survey revealed that the average length of a trip via public transport is 6 to 8 kilometres and the average travel time is 30 to 60 minutes. The consultants concluded that a successful public transport system in Pune must cover distances of more than 6 kilometres and must considerably shorten the average travel time in order to convince users of private vehicles to switch to public transport (Consult Team Bremen, BGS Ingenieurconsult International, 2007, 22-23).

The DPR also collates data on Pune's transport system collected in previous studies. The authors analyse key traffic challenges in Pune and they highlight the increasing gap between transport demand and public transport supply, assessing the growth rate of private transport as "alarming" (ibid., 24). The report also provides an overview of the expected social, economic and environmental impacts of a tramway project, distinguishing between the construction and operational phase. For example, it predicts positive effects on local employment during the construction phase while the local employment level may drop temporarily in the operational phase, when rickshaw drivers may lose customers. Another example is the reference to land acquisition which is a highly-contested topic in the growing cities of Pune and Pimpri Chinchwad. The DPR points out that a tramway would require very little land, only a total 52.5 hectares would be needed for depots, workshops and other facilities (ibid., 63-64). The study also assesses the environmental regulation affecting a tramway project (for more details see section 5.2.3.2).

Moreover the DPR provides detailed information on storm water drainage, roadbed methods and cross sections (ibid., 115-132). It proposes a tramway alignment development in two phases, with detailed information about the exact routes, and the number, distance and running time between stops and recommends that the tramway be constructed in eight steps (ibid., 138-179).

Based on their experiences of building tramway networks in Germany, the project partners CTB and BGS then set up an initial calculation of the cost of building and operating a tramway in Pune and Pimpri (the authors emphasise that a more precise calculation should be conducted once the preliminary design of the tramway is completed). Taking into account Indian market conditions and in consultation with Indian experts the DPR roughly estimates the cost of building a tramway in Pune by calculating 40% of the costs of setting up a corresponding tramway in Germany. The consultancy fees are also calculated according German standards. They are divided into approximately 70% work share conducted by Indian consultants with 25% of the standard fees for German consultants and approximately 30% work share for international experts for whom 100% of German standard fees are calculated (ibid., 182). A detailed breakdown of the project costs is attached in the DPR appendix (33-47).

In order to prevent an adverse "impact on project finance and time" the DPR includes ten pages of risk assessment in the areas of land acquisition; approvals by local authorities; consultant, contractor and staff expertise; material; climatic conditions; power supply; traffic projections; and potential interruptions (Consult Team Bremen, BGS Ingenieurconsult International, 2007, 208-218). However, like the cost calculation, the risk analysis appears to be largely based on the project consultants' experiences from building tramways in Germany and lacks a specific analysis of the Pune context.

In the DPR preamble the authors acknowledge that the "services to be provided comprise only preliminary and basic design & research work" (ibid.,) and in chapter 3 they highlight that more in-depth research has to be conducted in several areas such as social and economic risk and impact assessment in the city of Pune: "Necessarily backbone structures like land utilization, city and region characteristics, demographic and earning structures have to be investigated. Consequences of spatial, traffic and economic potentials and conflicts have to be evaluated." (ibid, 68-69).

Systematic Cooperation, Continuity of Interaction and External Moderator/Facilitator

Only during the phase when the three consultants jointly prepared the DPR did the tramway partnership project follow a systematic approach towards cooperation. Sakhalkar explains that the consultants shared the responsibilities according to their expertise (the German consultants were in charge of the technology and the urban transport planner brought knowledge of the local context and drew up the tramway alignment map) and that they collaborated smoothly and in a cooperative atmosphere. In July 2007 the consultants submitted the study to the city administrations of Pune and Pimpri Chinchwad, which marked the completion of the first step of the project (Interview with urban transport planner, 11-10-2013). The DPR provides detailed recommendations for the planned implementation. It suggests that the local authorities build the tramway in two phases, starting with six lanes encompassing a total of 91 km in the first phase. The authors recommend eight stages for each subsection of construction; beginning with the demolition or retrofitting of the existing infrastructure, followed by the construction of the track and road, the electricity supply and signalling equipment and finally the landscape design (Consult Team Bremen, BGS Ingenieurconsult International, 2007, 178-179).

However, the implementation of the DPR never began. According to the interviewed PMC official one reason for the project never being implemented was that the authorities and citizens in Pune lacked an understanding of and confidence in the tramway technology. He suggests that the project may have performed better if it had started with the construction of one demonstration lane to garner local support (Interview with PMC, 24-09-2013).

With regards to the continuity of interaction between the project partners the collaboration only performed well before the DPR was submitted. Representatives of the city administrations from Pune, Pimpri Chinchwad and Bremen conducted several exchange visits before they signed the tramway MoU in July 2005. During the preparation of the DPR the project partners Boltz, Monsees, Steiger and the urban transport planner also met regularly. They conducted four to five exchange visits each to Bremen and Pune respectively and they also frequently communicated via email and telephone (Interview with urban transport planner, 11-10-2013).

But the exchange between the German and Indian project partners ended rather abruptly when the DPR proposal was rejected by the Maharashtra state government and the remaining project funds were not paid by the Pune and Pimpri city administrations. The urban transport planner recalls that communication with his partners from Bremen was also hampered by the fact that Monsees left CTB after the submission of the DPR and that he had difficulties to keep in touch with the LAFEZ. The urban transport planner thus focused more on his collaboration with Steiger from Frankfurt with whom he started cooperating with on other projects (Interview with urban transport planner, 11-10-2013). Another interviewee from Pune confirms that the replacement of Monsees temporarily impeded the communication between Pune and Bremen which set the project back (Interview with NGO Arbutus, 25-09-2013).

As for the involvement of external support the Pune-Bremen partnership protagonists have generally preferred not to rely on outside help from external donors or government agencies. The partnership has been based on exchange between local individuals and institutions, which the partnership actors consider as more efficient and less bureaucratic (for more detailed information please refer the DEWATS case study in section 5.1). This approach is also reflected in the tramway initiative where the knowledge transfer in the DPR was conducted by a small team of actors without any external support and facilitation. The core project team was limited to the three consultants, Boltz and Hilliges from the Bremen city administration, plus Mahajani from the partnership organisation AFG (Interview with urban transport planner, 11-10-2013).

Policy Windows and Intrinsic Interests

Analysis of the political conditions that surrounded the tramway partnership initiative reveals that the project was not facilitated by a clear policy window of opportunity; a policy window was at best only partly and temporarily open. While the problem stream was rather favourable, the policy and political streams were less advantageous.

Similar to the DEWATS partnership project the challenge that the tramway system was to address was evident. Pune and Pimpri Chinchwad faced ever increasing traffic congestion and resulting air and noise pollution. The interviewed PMC representative highlights that everybody within the city administration agrees that there is a need to improve public transportation in Pune (Interview with PMC, 24-09-2013). The urban transport planner states that the Pune-Bremen partnership actors approached the Pune city administration with the idea of transferring the Bremen tramway to Pune, just when the city's decision makers were looking for an innovative public transport system to respond to its pressing traffic problems (Interview with urban transport planner, 11-10-2013). The DPR also mentions that the high demand for a mass transport solution in Pune and Pimpri had been voiced by many experts and in many previous studies. Due to Pune's narrow streets and the lack of land, a transport system based around the existing road infrastructure was required. The DPR concludes that a tramway would be the public transport system best suited to these conditions (Consult Team Bremen, BGS Ingenieurconsult International, 2007, 74).

The view that a tramway would actually improve Pune's traffic problems is however contested. A journalist from Pune points out that the sharing of narrow streets by both private vehicle and a tramway would not be feasible in the center of Pune as the traffic would make the running of the tramway very slow and passengers would be put at risk by having to embark or disembark from the trams in the middle of the street (Interview, 18-09-2013). Moreover, the tramway also competed with several other modes of public transport that were under discussion in the city at that time. The PMC representative explains that in addition to the tramway DPR the Pune city administration had commissioned proposals for a metro, a bus rapid transit (BRT) system, a sky bus and a mono rail in order to compare the benefits and shortcomings of each system. He adds that a Comprehensive Mobility Plan (CMP) was prepared to assess which technologies would fit best with the local context. The CMP recommended a multi-modal transport system, with a metro serving the city's high traffic density corridors, a tramway or sky bus for the medium density corridors, a BRT system for lower density roads and normal buses or rickshaws for the least frequented roads (Interview with PMC, 24-09-2013).

At the beginning of the collaboration the tramway was seen as a viable transport alternative by the local authorities, aided by the fact that the competing sky bus system was under pressure after an accident at a test facility in Goa. Pune's then Commissioner Nitin Kareer announced: "We (PMC and PCMC) are consulting German technical experts and we want a tram that suits our traffic conditions." (Indian Express, October 1st, 2004). But this support started to wane when local officials realised that the proposed tramway alignment also incorporated the highest density corridors, in particular the highway connecting Pune and Pimpri Chinchwad This road had been promoted by the CMP as the best route for a metro as well, which was prioritised by local decision makers. The urban transport planner explains that as metros gained ever greater political traction in India, the state government abolished its tramway plan (Interview with urban transport planner, 11-10-2013).

One key factor for the initial progress of the tramway project and its later failure has thus been the changing priorities of the local authorities in Pune and Pimpri Chinchwad. Initially the tramway initiative was designed to address the intrinsic interests of Pune, Pimpri Chinchwad and Bremen. The PMC and PCMC were interested in exploring and comparing different forms of public mass transportation and Pune's Commissioner Kareer considered the tramway a viable option to reduce the levels of traffic (Indian Express, October 1st, 2004). This interest corresponded with Bremen's attempt to strengthen business collaboration in the city partnership. Aside from the DEWATS initiative, supporters of the project also promoted the potential co-benefits of a tramway transfer for Bremen. The tramway project was set up as a business case and the initiative was an opportunity for companies such as the Bremen tramway company CTB to explore new markets (Interview with Interview with Forum Städtesolidarität Bremen-Pune, 01-10-2014). If the project had been successful, it would have been a source of employment in Bremen and proven to the local population that the city partnership with Pune actually provided concrete benefits for the local economy and workforce in Bremen (Interview with journalist from Pune, 18-09-2013). But when the tramway

DPR was rejected by the Maharashtra state government and the prospect of building a metro started to feature on the political agenda, the local authorities from Pune and Pimpri Chinchwad lost interest in the tramway system and no longer offered support for the partnership project.

5.2.3.2 Linkages between the Partnership Project and State Institutions

Similar to the DEWATS partnership project, the initiative to transfer the Bremen tramway to Pune and Pimpri Chinchwad was only marginally institutionalised into the state system. The involvement of local state institutions waned during the project and the transfer eventually failed due to the lack of support from the local and sub-national state authorities. While the Pune and Pimpri Chinchwad state authorities provided some funding for the project DPR, the project was largely coordinated and executed by non-state actors. The state leadership in the project was rather limited, as well (see table 9).

Indicators: State Institutionalisation	Evaluation: Tramway		
	Pune	Bremen	Total
Formal Approval by Local State Institutions	+-	+-	+-
Formal Approval by (Sub)National State Institutions	--	--	--
Public Human Resources	--	+-	+--
Public Financial Resources	+-	--	+--
Public Coordinator	--	+-	+--
Commitment of local State Leaders	+-	+-	+-

Table 9: Partnership Project Tramway: Findings from Index State Institutionalisation

Formal Approval by Local and (Sub)National State Institutions

The tramway project was legally obliged to build formal links to local and (sub)national state institutions as it required approval from local, state and national authorities in India for its implementation. The links to state institutions remained however weak and eventually the project failed as it did not receive approval from the Maharashtra state government.

The project was first discussed during meetings on the second partnership MoU in 2003. Along with the DEWATS project the tramway collaboration was among the first partnership initiatives that had the target of boosting the involvement of the local state authorities in the Pune-Bremen city cooperation. After several exchange visits by public officials between April 2003 and March 2005, a delegation of 45 representatives from Pune and Pimpri Chinchwad visited Bremen in July 2005 and signed a MoU to establish a tramway system connecting the two Indian cities in collaboration with the Bremen city administration (Forum Städtesolidarität Bremen-Pune e.V., 2006). The commissioners of Pune and Pimpri Chinchwad got financial approval from their cities' Standing Committees to jointly fund the project DPR and they delegated the coordination and execution of the report to the three consultants, Monsees, Steiger and the urban transport planner interviewed for this study (Interview with PMC, 24-09-2013).

While the cooperation between the consultants worked well, the project eventually failed when it again required approval from the state authorities for its implementation. The DPR highlights the need for the close coordination of the responsible state authorities to avoid delays and rising costs:

> Approvals of local authorities are necessary in case of track and road design, building and bridge calculations, water drainage and environmental impact. Missing or delayed approvals can provide a delay of start of the construction phase with an increase of the project costs, and any conditional approvals by the authorities lead to additional planning of buildings or infrastructure works. (Consult Team Bremen, BGS Ingenieurconsult International, 2007, 209)

While the local authorities' approval was required for decisions about the specific local design, the tramway first needed the general approval by higher policy institutions, i.e. the Maharashtra state government and by the central government in New Delhi who are the major decision-making units for large urban infrastructure projects in India. After the consultants submitted the DPR to the local authorities in Pune and Pimpri, they forwarded it to the Maharashtra State government for approval. The state government however rejected the proposal and afterwards it was not even handed over to the central government. The urban transport planner pointed out that the state and central governments were strongly promoting metro systems at that time and that they were not open to other public transport solutions, no matter how financially viable they were (Interview with urban transport planner, 11-10-2013).

Public Human and Financial Resources

The project DPR was coordinated and executed by a public-private project team, with a majority of private actors. On the side of Pune and Pimpri Chinchwad only non-state actors were actively involved in the preparation of the DPR. The urban transport planner worked as a private consultant in the project and Mahajani promoted the initiative as chairman of the AFG and head of the IOA 21. Both were engaged in the project voluntarily without being officially employed by the city administration (Interview with PMC, 24-09-2013).

On the German side the DPR was mainly executed by the private consultants Monsees from CTB and Steiger from the BGS. CTB is a subsidiary of the public tramway corporation Bremen Straßenbahn AG (BSAG) but it is registered as a private company. The BGS is a private "firm of planning and consulting engineers" (Consult Team Bremen, BGS Ingenieurconsult International, 2007, 2) that is part of the European Grontmij Group. The only state actors were Boltz and Hilliges from the Bremen city administration who helped initiate the project.

The Pune and Pimpri Chinchwad city administrations had agreed to cover in full the 50,000 Euro cost of hiring the two German consultants Monsees and Steiger to prepare the tramway DPR. The PMC however refused to pay the second half of the stipulated amount. As a consequence, the indicator measuring the state funding for the partnership project is only partly fulfilled as the private consultants had to bear a share of the costs for the preparation of the project DPR from their own companies' budgets.

Public Coordinator and Commitment of Local State Leaders

Local state institutions' influence was even more limited in the coordination of the tramway initiative. The project was initiated and driven by the two partnership entrepreneurs, Hilliges and Mahajani. Hilliges was employed by the LAFEZ in the Bremen city administration, but he also invested his private time and resources in the Pune-Bremen partnership (Interview with Interview with Forum Städtesolidarität Bremen-Pune, 01-10-2014). Mahajani was heading the publicly funded IOA 21, but he was not employed by the Pune city administration and worked largely voluntarily at the office, promoting the tramway project for idealistic reasons (see the following section 5.2.3.3 for more details about the two partnership entrepreneurs). The DPR does not mention any involvement of state actors in the coordination of the study, highlighting that the report was conducted in cooperation with CTB and BGS engineers (Consult Team Bremen, BGS Ingenieurconsult International, 2007, 2-3). The PMC representative adds that the private local transport planner from Pimpri Chinchwad oversaw the DPR preparation on the side of Pune and Pimpri Chinchwad (Interview with PMC, 24-09-2013).

The local public decision makers from Pune and Pimpri Chinchwad only committed to the tramway initiative at the beginning of the project. The PMC representative reports that the then Pune Commissioner Pravinsinh Pardeshi quickly approved the DPR proposal as he was a former UN representative who worked in Switzerland and was familiar with city development in Europe (Interview with PMC, 24-09-2013). According to this interviewee the then commissioner of Pimpri Chinchwad, Aseem Gupta, was also very supportive of the tramway project. Gupta had relations in Bremen and had visited Bremen before (ibid.). The interviewee from Bremen's partnership association confirms that the personal commitment of local leaders is crucial for larger partnership projects and that at the outset of the tramway project a team of very engaged mayors and commissioners supported the Pune-Bremen partnership activities. He adds that they visited Bremen for a test ride in the tramway and showed interest in transferring the technology. However, they were replaced shortly afterwards and their successors were less committed to building a tramway in Pune while they were pushing for a metro (Interview with Interview with Forum Städtesolidarität Bremen-Pune, 27-02-2013).

Also the commitment of the political decision makers from Bremen to the tramway transfer initiative was mixed. The then mayor of Bremen, Scherf, generally favoured the tramway initiative (Interviews with urban transport planner, 11-10-2013 and Forum Städtesolidarität Bremen-Pune, 01-10-2014). However, several political actors from Bremen, in particular the State Councillor of the State Department for Building, Environment and Transport, were sceptical of the initiative as they were concerned about the risks of such a large-scale project. Only Boltz who worked in the State Department for Building, Environment and Transport strongly supported the tramway project, even travelling to Pune to support the preparation of the DPR at his own expense (Interview with Forum Städtesolidarität Bremen-Pune 01-10-2014).

5.2.3.3 Local Partnership Entrepreneurs

The tramway partnership project was initiated and driven by the two partnership entrepreneurs, Hilliges from Bremen and Mahajani from Pune who were also instrumental in fostering the DEWATS project (see prior case study). They score similarly high in the tramway index system with Hilliges again fulfilling all six indicators for a successful partnership entrepreneur (see table 10).

Indicators: Local Partnership Entrepreneur	Evaluation: Tramway		
	Pune	Bremen	Total
Belief in Feasibility and Benefits	++	++	++
Investment of own Resources	++	++	++
Ability to Convince Stakeholders	+-	++	++-
Internal Policy Network	+-	++	++-
External Policy Network	+-	++	++-
Partner City Policy Network	+-	++	++-

Table 10: Partnership Project Tramway: Findings from Index Partnership Entrepreneur

Belief in Feasibility and Benefits, Investment of Resources and Ability to Convince Stakeholders

Both Mahajani and Hilliges were convinced of the benefits of transferring the Bremen tramway system to Pune. In particular Mahajani pursued the vision of transferring the Bremen tramway system to Pune with considerable persistence, as pointed out by the DEWATS project advisor who worked with Mahajani at the IOA 21. Mahajani had admired the tramway network on numerous visits to Bremen and it was one of his aims in life to establish a similar system in Pune (Interview with DEWATS project advisor, 08-10-2013). The interviewed PMC representative, then working for the transport department of Pune Municipal Corporation, repeatedly communicated with Mahajani about the tramway project and confirms that Mahajani was convinced that a tramway could give Pune's public transport system a boost and help improve quality of life in the city (Interview with PMC, 24-09-2013). The urban transport planner states that Mahajani took personal ownership of the project and was instrumental in the realisation of the DPR (Interview with urban transport planner, 11-10-2013). In the DPR, Mahajani's engagement for the partnership project is underlined by the authors who extend their "special thanks" to Mahajani, "who was strongly committed to ensure the cooperation needed for the project" (Consult Team Bremen, BGS Ingenieurconsult International, 2007, 3). The PMC representative states that both Mahajani and Hilliges took personal ownership and responsibility for the progress of the initiative (Interview with PMC, 24-09-2013).

Hilliges was also convinced of the benefits of building a tramway in Pune. Amongst others he organised a test ride in the Bremen tramway for a visiting delegation of members of the Pune and Pimpri local authorities in order to convince them of the benefits of the tramway system and was very disappointed when the transfer eventually failed (Interview with Forum Städtesolidarität Bremen-Pune, 27-02-2013). On several of his visits to Pune Hilliges also helped Mahajani to promote the tramway (Interview with journalist from Pune, 18-09-2013). Even five years after the tram initiative had been rejected by the Maharashtra state government, Hilliges tried again to find out what had happened to the tramway DPR and whether the local authorities in Pune and Pimpri were still interested in the report. He asked the head of the NGO Arbutus who had taken over leadership of the partnership after Mahajani passed away to contact the Pune city administration and inquire about the study. The head of Arbutus met twice with a PMC representative who had been involved in the tramway project but this PMC employee referred him to the Pimpri Chinchwad city administration which he said was now responsible for the DPR. The head of Arbutus then opted not to follow-up on this due to personal time constraints (Interview with NGO Arbutus, 25-09-2013).

As elaborated in the DEWATS case study, both Hilliges and Mahajani spent substantial amounts of their own financial and time resources on the Pune-Bremen partnership (see section 5.1.4.3). The DEWATS project advisor highlights that Mahajani even risked his own personal reputation for the tramway project by continuing to fight for the tramway in Pune despite considerable scepticism in both Bremen and Pune (Interview with DEWATS project advisor, 08-10-2013).

Both Mahajani and Hilliges are described by the majority of the interview partners in this study as charismatic and respected citizens who passionately advocated the Pune-Bremen partnership in both cities. Mahajani is praised as an honourable and "genuine person" who had the ability and personal attitude to talk to different audiences regardless of their rank (see section 5.1.4.3). The DEWATS project advisor had however the impression that while Mahajani was generally a respected figure in Pune, he was not always taken seriously by the Pune city administration when he continued to promote the tramway project (Interview with DEWATS project advisor, 08-10-2013).

Hilliges has been characterised as enthusiastic and dynamic, investing great coordinating efforts into the Pune-Bremen partnership (see section 5.1.4.3). The urban transport planner from Pimpri Chinchwad remembers that Hilliges was a convincing advocate for the tramway initiative as he had a lot of experience in India (Interview urban transport planner, 11-10-2013). This interviewee adds that Hilliges is charismatic, able to engage with different groups of people and that he was thus instrumental in pushing the project during its initial phase (ibid.). The representative of Arbutus confirms that Hilliges made great efforts to convince local decision makers from Pune and Pimpri Chinchwad of the benefits of the Bremen tramway and that he managed to raise their interest (Interview with NGO Arbutus, 25-09-2013).

Internal, External and Partner City Network

Hilliges' and Mahajani's access to policy networks in and beyond Bremen and Pune has been elaborated in detail in the DEWATS project case study (see section 5.1.4.3); it is also summarised in this section with the addition of specific references to the tramway project.

Both Hilliges and Mahajani had close contact to the local policy networks in their respective cities. Hilliges strongly benefited from his professional position as head of the LAFEZ in the Bremen city administration which enabled him to directly contact key officials and decision makers through official channels. He also has good connections with local NGOs and companies in Bremen. In the tramway initiative Hilliges facilitated the addition of consultant Monsees from the Consult Team Bremen to the project team that prepared the DPR.

While Mahajani's access to local NGOs was rather limited, he had developed good and sometimes even personal relations with key local political and industry leaders. Together with the DEWATS project the tramway initiative was one of the few attempts to boost the involvement of local political actors in the partnership activities. Mahajani was instrumental in initiating contact with the commissioners of Pune and Pimpri Chinchwad at the beginning of the project and Mahajani was announced by the Pune authorities as their key local partner in the tramway initiative (Indian Express, October 1st, 2004). The partnership's dependency on Mahajani for access to local politicians became apparent when the partnership protagonists lost contact with the Pune Municipal Corporation after Mahajani passed away (Interview with Forum Städtesolidarität Bremen-Pune, 27-02-2013).

Mahajani's political contacts outside of Pune were generally less extensive which meant that he was not able to mobilise support at higher policy levels after the tramway DPR was sent to the Maharashtra state government. Hilliges, as the official representative of the Bremen city administration, head of Terre des Hommes Germany and co-founder of the international initiative "Towns and Development", had established comprehensive relations outside of Bremen, particularly in the area of development cooperation. However Hilliges' external network did not directly benefit the tramway project.

With regard to contacts within the respective partner cities Hilliges had established a network of close contacts to a group of citizens and several local NGOs in Pune that he has visited numerous times over the past decades. Hilliges even developed friendships with several partnership actors, amongst them Mahajani.

Hilliges also had regular access to local politicians in Pune via Mahajani. Interviewees highlight that both in the tramway project as well as in other partnership activities Hilliges has always been the key contact person for anyone in Pune who was interested in the Pune-Bremen partnership (Interview with urban transport planner, 11-10-2013; Interview with PMC, 24-09-2013).

Mahajani visited Bremen less frequently than Hilliges did Pune and he relied in large part on Hilliges to involve partners from Bremen in joint exchange activities as part of the city cooperation (Interview with DEWATS project advisor, 11-09-2014).

5.2.3.4 Partnership Social Capital

With regard to the building of social capital between the partnership protagonists from Pune and Bremen the tramway initiative achieves a medium score in the index system of this study. The project stands out positively thanks to its experiences from prior partnership activities, but it scores lower on inclusion and citizen participation. In the remaining four indicators for partnership social capital the tramway initiative gained mixed results (see table 11).

Indicators: Partnership Social Capital	Evaluation: Tramway		
	Pune	Bremen	Total
Prior Collective Action	/	/	++
Trust	+-	+-	+-
(Perceived) Equality	+-	+-	+-
Intercultural Experiences	+-	+-	+-
Intercultural Communication	/	/	+-
Involvement of Leadership Networks	+-	--	+--
Inclusion & Citizen Participation	--	--	--

Table 11: Partnership Project Tramway: Findings from Index Partnership Social Capital

Prior Collective Action, Trust and (Perceived) Equality

As in the DEWATS project, the tramway initiative could draw on extensive experiences from prior partnership activities in sustainable development and other areas that had been conducted since the mid-1970s as part of the Pune-Bremen cooperation (see sections 5.1.1 and 5.1.4.4 for more details about the history and broad range of activities that have been organised as part of the Pune-Bremen partnership). Public transport has become a partnership topic during the Local Agenda 21 cooperation in the 1990s. Before the tramway collaboration Bremen had already supported Pune in projects to improve the environmental performance of its public bus system; Bremen had offered training on eco-efficient transportation for bus drivers from Pune and provided nozzle testing equipment for the public bus depots to control and reduce exhaust emissions (Forum Städtesolidarität Bremen-Pune e.V., 2006).

The member of the Forum Städtesolidarität Bremen-Pune points out that the prior partnership experiences, particularly the public bus project, helped build a trusting atmosphere during the first years of the tramway cooperation (Interview with Forum Städtesolidarität Bremen-Pune, 01-10-2014). The interviewee from PMC confirms that relations between the partnership actors from Pune and Bremen were very good when the tramway project was initiated. He adds that in particular the two partnership entrepreneurs, Hilliges and Mahajani, were able to build trust among the project partners as they had an understanding of the cultural differences between India and Germany (Interview with PMC, 24-09-2013). Also the urban transport planner confirms that Hilliges and Mahajani built the ground for a trusting collaboration between the project consultants in the preparation of the DPR (Interview with urban transport planner, 11-10-2013). The DPR emphasises the high degree of trust between the project partners, as well: "The cooperation of Pune and Bremen within their city sistership enables an intensive exchange of knowledge and ideas, which contributes to a very trusting discussion to satisfy the rapidly growing traffic demand of Pune region in a sustainable and environment friendly way." (Consult Team Bremen, BGS Ingenieurconsult International, 2007, 2).

The actors from Bremen and Pune however lost confidence in each other's reliability and credibility after the DPR was rejected by the Maharashtra state government. The interview partners pointed to two events leading to the loss of trust in the project; firstly, the refusal of the PMC to transfer the second instalment of the DPR funding (Interview with Forum Städtesolidarität Bremen-Pune, 27-02-2013), and secondly, the withdrawal of the project consultant, Monsees, from the project (Interview with NGO Arbutus, 25-09-2013).

As already elaborated in the DEWATS case study the Pune-Bremen partnership, like many development cooperation initiatives, has struggled to achieve real equality among the protagonists. A key challenge for building equal partnerships derives from the one-sided funding of joint projects, usually provided by the partner from the Global North (see section 5.1.4.4). The tramway initiative was an attempt to reduce this inequality by setting up a business-oriented partnership project which was to be financed by the partner from the Global South, the city of Pune, and supported by a joint team of German and Indian private consultants. In the preparation of the DPR the consultants Monsees, Steiger and the urban transport planner from Pimpri Chinchwad cooperated well and on the same level, with each consultant taking over distinct roles. The German consultants oversaw the technology while the urban transport planner had responsibility for the setting up of the tramway alignment plan and feeding information about the local context conditions into the project (Interview with urban transport planner, 11-10-2013). The Pune and Pimpri Chinchwad city authorities also initially agreed to fully fund the DPR. However, they eventually refused to pay the second instalment of the stipulated sum. Furthermore, an interviewee points out that enthusiasm for the tramway remained rather one-sided, as the PMC did not make much effort in the preparation and development of the project: (Interview with DEWATS project advisor, 08-10-2013).

Intercultural Competence and Communication

As for the partnership actors' intercultural competences, two of the protagonists that were directly involved in the preparation of the tramway DPR study had already worked on international projects before. Boltz had conducted several projects in Eastern Europe and Monsees had worked in developing countries. For all three project advisors from Germany (Boltz, Steiger and Monsees) it was however the first project they had worked on in India. The urban transport planner from Pimpri Chinchwad had also never visited Germany prior to the tramway project (Interviews with urban transport planner, 11-10-2013 and Forum Städtesolidarität Bremen-Pune, 01-10-2014). The two involved partnership entrepreneurs, Hilliges and Mahajani, were very familiar with India and Germany, respectively, due to Mahajani's studies in Germany and a multiplicity of exchange visits and projects as part of the long-term city partnership between Pune and Bremen.

Despite limited experiences of the partner countries there were no major language and communication problems among the actors directly involved in preparing the DPR. The project consultants and the two partnership entrepreneurs were all fluent in English and Mahajani even spoke both English and German (Interview with urban transport planner, 11-10-2013). Language barriers did however play a role in the project with regard to the communication with other actors from Pune and Bremen. The urban transport planner points out that, other than the younger generation today, many people in Bremen were not able to communicate fluently in English at the time of the tramway project's initiation in 2005 (ibid..). Interviewees from Bremen state that Pune's Mayor Chaudhari who signed the tramway MoU, was well educated and fluent in English but her successors did not speak English which made interaction difficult (Interview with Forum Städtesolidarität Bremen-Pune and LAFEZ, 27-02-2013).

Involvement of Leadership Networks, Inclusion, and Citizen Participation

The tramway initiative scores lower in the remaining partnership social capital indicators measuring the involvement of local informal leadership networks and inclusion.

According to the interviewed urban transport planner, apart from the project consultants, the two partnership entrepreneurs and Boltz from the Bremen city administration, no local stakeholders from Pune or Bremen participated in the preparation of the tramway DPR (Interview with urban transport planner, 11-10-2013). The PMC representative explains that they had planned to involve local stakeholders from Pune and Pimpri Chinchwad once the project had been launched. But he also points out that there is generally a rather negative atmosphere between the city administration and local NGO activists (Interview with PMC, 24-09-2013). The journalist from Pune confirms that key local NGOs dealing with transportation challenges in Pune did not engage for the tramway project (Interview with journalist from Pune, 18-09-2013).

The only reference to local stakeholder involvement can be found in the project DPR which mentions that local industries have voiced their interest in abandoning their private bus services to the industrial facilities outside of Pune in case the tramway was build: "The Industries have come forward with a strong willingness to phase out their existing private transport system (fleet of private company busses) after introduction of the Tramways." (Consult Team Bremen, BGS Ingenieurconsult International, 2007, 140). The DPR calculates that at least 1000 private buses could thereby be removed from Pune's and Pimpri Chinchwad's streets (ibid., 222).

The DPR includes few references to the inclusion of underprivileged sections of the Pune population in the tramway project. It mentions in the introduction that slums are a major challenge in Pune (ibid., 18) without detailing how slums will be (positively or negatively) affected by tramway. The report only states later that the tramway implementation would not cause additional exclusion through the displacement of citizens as the construction of the tramway required little additional land (ibid., 64).

5.3 Case Study Nashik – Hamburg: Reducing Emissions through Waste-to-Energy

5.3.1 Partnership Project W2E: Content and Process

Since 2009, the Nashik Municipal Corporation (Nashik MC[46]) and Hamburg's public water utility, Hamburg Wasser, have collaborated in the construction of an innovative waste-to-energy (W2E) plant in the western Indian city of Nashik. The partnership has been facilitated and coordinated by the Deutsche Gesellschaft für Internationale Zusammenarbeit GmbH (GIZ), the German governmental organisation for development cooperation. Compared with the Pune-Bremen partnership projects, the exchange between Nashik and Hamburg exemplifies a more formalised, top-down initiated and predominantly public approach to transnational urban climate cooperation.

The aim of the collaboration is to transfer the Hamburg Water Cycle® (HWC®), a waste-to-energy technology which was developed by Hamburg Wasser, to Nashik. Hamburg Wasser has tailored the HWC® to the special requirements of fast growing cities worldwide that are searching for alternative sanitation solutions to conventional centralised wastewater treatment. The HWC® has been designed to close material loops in wastewater treatment to save water, energy and carbon emissions. Its special features include the decentralised separation of grey and concentrated black water streams at source, plus the recycling of the waste's energy and nutrients. The HWC® technology is very flexible and can be adapted to specific urban context conditions such as water shortages, periods of regular heavy rainfall and high or low availability of organic waste (Hamburg Wasser Kompetenznetzwerk, XVI-XXI).

For the project in Nashik, Hamburg Wasser adjusted the HWC® to meet the demands of densely populated and expanding cities in India that rely on the inefficient and costly disposal of black water from communal toilets and organic kitchen waste. In consultation with the GIZ, Hamburg Wasser adapted the HWC® to simultaneously treat organic solid waste and black water in an eco- and climate-friendly way. Prior to the project in Nashik Hamburg Wasser had only tested this form of co-fermentation at the conceptual level and it aims to use the plant in

---

46 Officially, Nashik Municipal Corporation and Nagpur Municipal Corporation (see following case study) are both abbreviated 'NMC'. For the purpose of differentiation they are abbreviated 'Nashik MC' and 'Nagpur MC' in this study.

Nashik as a pilot facility to demonstrate the feasibility of this novel approach (ibid., XXI-XXIII; interview with Hamburg Wasser, 01-03-2013). A GIZ employee points out that the waste-to-energy plant in Nashik will be the first ever combined organic waste and black water treatment facility in India (Interview with GIZ, 05-12-2012).

The plant is able process 31 metric tons of kitchen waste from local hotels and restaurants together with septic waste from municipal toilets daily. It treats the organic waste of almost 100% of local hotels and restaurants and about 60-70% of the black water from public toilets (Interview with GIZ, 05-12-2012). Through anaerobic processing of the waste, the plant will produce up to 3200 kWh of renewable electricity per day and avoid uncontrolled methane emissions. Thereby a total of 4,700 tons of CO_2-equivalent emissions are prevented every year. Additionally, in the co-fermentation process all nutrients are recovered in order to produce a replacement for the artificial fertilizers that are currently used by local farmers (Augustin, Giese and Dube, 2010).

The main project partners in this initiative are the GIZ, Nashik MC, Hamburg Wasser and its subsidiary Consulaqua Hamburg, the Indian consultant Paradigm Environmental Strategies Pvt. Ltd. and the private contractor who builds and runs the waste-to-energy plant.

The concept for the project was developed by a GIZ employee working under the Indo-German Environment Partnership (IGEP) in New Delhi. The GIZ-IGEP has been supporting the urban local bodies in Nashik and six other Indian cities (Shimla, Varanasi, Raipur, Kochi, Nainitel and Tirupathi) in implementing the National Urban Sanitation Policy by providing technical assistance and consulting services in sustainable urban sanitation (Hamburg Wasser Kompetenznetzwerk, 2010, 1). Within this framework, the GIZ has initiated and organised the collaboration to build a waste-to-energy plant in Nashik. The GIZ serves as the formal project coordinator and is directly responsible for the project's outcome to the funder, the German Federal Ministry for the Environment, Nature Conservation and Nuclear Safety (BMU[47]). The GIZ involved all other partners in the project and it also facilitates and controls all communication between the project partners (Interview with GIZ, 27-11-2013). In the project's Memorandum of Understanding signed by the GIZ and Nashik MC, the GIZ agrees to engage external experts to support the preparation, construction, operation and monitoring of the waste-to-energy plant. Furthermore, the GIZ confirms that it will support Nashik MC in selecting a private operator; providing technical assistance and capacity building for Nashik MC and operator staff; and raising external funds to cover the capital

---

47 In 2013, the BMU was renamed in 'Federal Ministry for the Environment, Nature Conservation, Building and Nuclear Safety' (BMUB).

costs (Memorandum of Understanding between GTZ-ASEM and Nashik Municipal Corporation, 2010). Nashik MC is responsible for the construction of the pilot plant and for organising and guaranteeing the waste collection infrastructure as well as the availability of sufficient waste to run the plant. Moreover Nashik MC commits to assigning a nodal officer and experienced staff for the operation and maintenance of the project and to provide the land for the facility. Nashik MC also agrees to support the GIZ in securing legal approval for the project (ibid.).

Hamburg Wasser's main contribution to the project is to provide the HWC® technology and to support the micro planning process of the waste-to-energy plant in Nashik. The utility prepared the feasibility study for the project and facilitated planning documents such as the project DPR and the tender. During the construction phase Hamburg Wasser's private subsidiary Consulaqua Hamburg which specialises in international project implementation will take over the responsibility for the monitoring of the project; Consulaqua Hamburg will inspect the tender and the construction process and has the right of objection with regard to technical and financial aspects of the project (Interview with GIZ, 27-11-2013). In addition, Consulaqua has been commissioned to support the GIZ in training the personnel working on the project and to develop maintenance and environmental protection plans (Interview with Hamburg Wasser, 01-03-2013).

The GIZ also hired a private Indian consultancy, Paradigm Environmental Strategies Pvt. Ltd. in Bangalore, to prepare the project DRP and help Nashik MC and Hamburg Wasser in the compilation of the tendering documents. According to a Hamburg Wasser interviewee involved, Paradigm was also a crucial partner in the financial calculations, providing Hamburg Wasser with information on which materials and equipment could be purchased from local Indian sources (Interview with Hamburg Wasser, 01-03-2013).

In consultation with Hamburg Wasser and Nashik MC, the GIZ decided on a public private partnership model for the execution and financing of the waste-to-energy plant in Nashik. Another important project partner is therefore the private contractor that will be selected during the tendering process and that will take on the responsibility for the additional funding required, the specific design and construction, and the operation and maintenance of the plant (GIZ, Write up on Waste to Energy Project for Standing Committee).

The majority of the project investment costs are covered by the BMU's funding program "International Climate Initiative" (ICI). Since 2008 the ICI has funded climate protection projects in developing and emerging economies amounting to 120 million Euros per year. These funds initially stemmed from European Union Emissions Trading System revenues and then were later covered by the BMU's budget[48]. The ICI in particular aims to foster international cooperation and technology transfer that goes beyond the formal international climate negotiations (Interview with GIZ, 27-11-2013).

The GIZ successfully applied for ICI funding for the setting up a waste-to-energy plant in one of the GIZ's partner cities in India and as part of this the GIZ conducted a pre-feasibility study on conditions linked to the success and failure of existing waste-to-energy plants in India. In January 2010 the GIZ invited Hamburg Wasser to suggest a technical proposal for the construction of waste-to-energy plants in India and hand in a cost calculation for the preparation of a feasibility study for the project (Hamburg Wasser Kompetenznetzwerk, 2010, 1). The GIZ hired Hamburg Wasser as a consultant for the project due to the utility's experience working with waste to energy systems and developing the HWC® technology that the GIZ assessed as suitable for the Indian urban context.

Initially, the GIZ were looking to build the waste-to-energy plant in Delhi and Hamburg Wasser employees visited Delhi to promote the HWC® technology and assess the local conditions for the project. It turned out that these conditions were less favourable than expected because in Delhi the organic waste required for a plant was already being utilised for pig farming. In consultation with Hamburg Wasser the GIZ then selected Nashik as the Indian partner city for this project. Nashik was chosen as it had sufficient organic material available and the city had already established the waste collection infrastructure required for the project (Augustin, Giese, Dube, 2010). A GIZ employee involved adds that another reason for selecting Nashik was the city's experiences in implementing larger scale projects, plus the fact that Hamburg and Nashik have roughly the same population (Interview with GIZ, 07-12-2012).

In April 2010 a Hamburg Wasser employee visited Nashik to collect the data required for the preparation of the feasibility study. In the same month the ICI Technical Co-operation Agreement was signed between the central governments of India and Germany (Dube, 2013) and on September 9th, 2010 the GIZ and the commissioner of the Nashik Municipal Corporation signed a Memorandum for Understanding for German-Indian collaboration in the setting up of a waste-to-energy plant in Nashik (Memorandum of Understanding between the GTZ-ASEM and Nashik Municipal Corporation).

In September 2010 Hamburg Wasser also completed its feasibility study. The authors concluded that a transfer of the HWC® technology to city context of Nashik would be feasible (Hamburg Wasser Kompetenznetzwerk, 2010, I). One month later Hamburg Wasser employees presented the feasibility study's findings and discussed the next project steps with officials from Nashik MC and the local

---

48 http://www.international-climate-initiative.com/en/about-the-iki/iki-funding-instrument/ (19-02-2016)

council at a GIZ workshop on waste management at the Nashik Local Centre (GTZ, 2010).

Over the following year the Indian consultant Paradigm prepared the Detailed Project Report (DPR) for the waste-to-energy plant in Nashik, based on the HWC® technology and submitted it to the GIZ and the Nashik MC in September 2011 (Paradigm Environmental Strategies Pvt. Ltd., 2011). In the same month, the Superintendent Engineer of Nashik Municipal Corporation who is in charge of the project's execution in Nashik, visited Hamburg to present the waste-to-energy project and the Nashik City Sanitation Plan at the Internationaler Umweltrechtstag, an international environmental law conference in Hamburg (Pawar, 2011). During his stay he also met the Hamburg Wasser consultants involved who introduced him to Hamburg Wasser's work and the HWC® approach (Interview with Hamburg Wasser, 01-03-2013).

In September 2011 the Indian Federal Ministry of Environment and Forest and the GIZ also signed the implementation agreement for the waste-to-energy plant in Nashik. According to the schedule the commissioning of the plant was planned for January 2014. However, the project suffered delays as it took Nashik Local Council more than a year to approve the financial plan, finally doing so in summer 2013. A GIZ employee involved states that receiving financial approval was a decisive step in ensuring the project's progress (Interview with GIZ, 27-11-2013).

At the time of the completion of the data collection for this study in Nashik (November 2013), the construction of the plant was scheduled for 2014 and its commissioning planned for early 2015 (Interview with GIZ, 30-09-2013). In a final telephone interview in September 2014 a GIZ employee involved that the project's implementation had been delayed once again because of an insufficient number of bidders during the first two tendering rounds. During the third round the project finally received enough bids and a contractor was selected. Due to local elections in May 2014 the administrative process was further held up, but according to this interviewee no obstacles now remain and the GIZ expects that the construction of the plant will eventually start in December 2014 (Interview with GIZ, 08-09-2014).

5.3.2 Partnership Project W2E: Outcomes

In the project outcome evaluation the partnership between the GIZ, Hamburg and Nashik to implement a waste-to-energy plant in Nashik scores high with regard to the (likely) project implementation as well as for realising partnership mutuality. It has achieved partial success in the areas of capacity building, benefits for the local target groups and the project's wider impact. The project outcome is assessed less positive with regard to the budget and schedule performance and the partnership post-project sustainability (see table 12).

Except for the indicator measuring the budget and schedule performance (which can already be assessed to score low), the scoring of the success indicators is however temporary, as the project is still ongoing. The outcome evaluation of the project is therefore tentative and the scoring may change in a later evaluation.

Success Indicators	Project Evaluation: Waste to Energy (tentative)		
	Nashik	Hamburg	Total
Implementation Achievements	/	/	++
Budget and Schedule Performance	/	/	--
Local Capacity Building	+-	+-	+-
Benefits for Target Group	/	/	+-
Impact	/	/	+-
Mutuality	/	/	++
Post-project Sustainability	/	/	--

Table 12: Partnership Project Waste to Energy: Findings from success index

Implementation Achievements

The W2E partnership project scores highly with regard to implementation achievements, despite the plant having not yet been constructed at the point of this study's completion. All preparatory work has however been finalised with the project partners conducting extensive planning and completing three in-depth studies in preparation for the project implementation (the pre-feasibility study, the feasibility study and the DPR). Furthermore, despite several delays the project has received all the major approvals required from the central and local governments and no administrative hurdles remain.

Budget and Schedule Performance

The W2E partnership project has suffered several delays. As a consequence, the project's timeline had to be extended several times. The Hamburg Wasser project consultants state that the project's completion is long overdue. They point out that according to the initial schedule the plant should have started operating by 2012 at the latest (Interview with Hamburg Wasser, 01-03-2013). The Hamburg Wasser employees were not aware of the latest adjustments to the project's schedule. They just knew that the project had been delayed since Consulaqua was commissioned to take over the monitoring of the project and that there had been no progress in 2012 (Interview with Hamburg Wasser, 01-03-2013).

The project MoU between the GIZ and Nashik MC signed on September 9th, 2010 schedules the project's completion for December 31, 2012 "at the latest" (Memorandum of Understanding between GTZ-ASEM and Nashik Municipal Corporation, 2010, 1). But approval from the local council was postponed and in an interview in September 2013, a GIZ employee involved stated that according to the latest plans the plant should be fully operational in about 18 months or by summer 2015 (Interview with GIZ, 30-09-2013 (2)). In a final telephone interview in September 2014 this employee highlighted that the project was encountering additional delays, delaying the commissioning of the plant even further (Interview with GIZ, 08-09-2014).

As a result of the delays the project subsidy that had been budgeted for the capital costs of the plant had to be reduced as the GIZ needed to cover its ongoing employment and administrative costs (Interview with GIZ, 27-11-2013).

Local Capacity Building

The GIZ sees capacity building as a crucial element in the realisation of sustainable development and pursues this in the majority of its partnership projects in developing and emerging economies. The GIZ focuses on the development of competences at individual, organisational and societal levels in order to enable the collaborating partners from the target countries to articulate, negotiate and implement their ideas of sustainable development (Stockmann, Menzel and Nuscheler, 2010, referring to Borrmann 2009: 90). A Senior Advisor at the GIZ Indian office in New Delhi, who co-supervises the waste to energy project in Nashik, points out that institutional and governance reforms as well as capacity building are crucial to establishing innovative sanitation solutions in India (Interview with GIZ, 27-11-2013). He points to the issue of Western countries having introduced centralised sewage networks many decades ago, so know-how and practical experiences on how to transform more traditional sanitation systems are very limited. As adequate larger-scale sewage solutions are in high demand in India but not yet available on the market, the GIZ aims to foster the development of innovative sanitation approaches that fit the Indian context (ibid.). In addition to Hamburg Wasser the GIZ cooperates with the Indian Goa Birla Institute of Technology and Science. This institute has links to Professor Otterpohl at the Technical University of Hamburg Harburg who has supported Hamburg Wasser in the development of the Hamburg Water Cycle®. The Birla Institute has already incorporated the HWC® technology into their academic portfolio and has conducted several studies on how to optimise biogas production. It will also support Nashik MC and the contractor in the construction and operation of the waste-to-energy plant in Nashik (ibid.).

A major reason for selecting Nashik as the site for the waste-to-energy plant was that the city already had some of the required capacities, such as the separate collection of waste streams, plus well-trained staff that have experience of running a municipal solid waste (MSW) plant. The feasibility study concludes that as "Nashik is the only city in Maharashtra which has taken lead towards scientific management of MSW in abidance of MSW rules 2000", Nashik has the potential to become a "lime-light training and development centre for the State of Maharashtra" (Hamburg Wasser Kompetenznetzwerk, 7). For the waste-to-energy plant the authors suggest the compilation of a detailed training concept for the operational staff highlighting that this is "at least of the same importance as the design [of the plant] and needs to be well adapted to the local conditions" (ibid., 27). The preparation should include on-the-job training "at a wastewater treatment plant (WWTP) with anaerobic digestion and biogas production for example in Germany" (ibid.), to be conducted after the construction of the plant and a few weeks prior to the commissioning. The start-up phase should be accompanied by experienced international experts to train the operational staff in operational, maintenance and safety issues. The study also recommends that a GIZ engineer provide training on waste and sewage collection (ibid., 27-28). GIZ and Hamburg Wasser employees confirm in their interviews that comprehensive training for Nashik MC and the contractor's staff are planned (Interviews with GIZ, 27-11-2013 and Hamburg Wasser, 01-03-2013).

Capacity building was also a key reason for Hamburg Wasser joining the partnership project. At the waste-to-energy plant in Nashik, Hamburg Wasser will test the simultaneous treatment of solid organic waste and black water as part of the HWC® approach for the first time. The Hamburg Wasser consultants emphasise that they aim to learn how the HWC® works in practise in India. They also plan to bring officials from German cities to Nashik to demonstrate the benefits of the combined treatment. One consultant pointed out that German cities often lack the capacity and will to conduct simultaneous waste and wastewater treatment. He hopes to initiate a move towards more integrated approaches in Germany. He explained that the fact that in Germany the sectors of waste and water management are often organised and regulated separately which hampers innovative trans-sectoral approaches. He believes that by demonstrating the benefits of integrated approaches in other countries it may be possible to help foster the required legal reforms for more integrated planning in German cities as well (Interview with Hamburg Wasser, 01-03-2013).

Benefits for Target Groups

The major local target group is the slum population in Nashik; they use public toilets as they lack private sanitation facilities. A GIZ employee involved highlights that as part of the implementation of the waste-to-energy project the cleanliness and hygiene of the public toilets will be enhanced. He points out that during their visit to Nashik for the workshop on October 19, 2010, the Hamburg Wasser consultants gave some recommendations on how to improve the hygiene of the septic tanks. These recommendations have already been implemented by local officials and public representatives (Interview with GIZ, 07-12-2012).

The GIZ employee explains that the waste-to-energy plant will bring improvements for the city administration as it will reduce the pressure on the existing waste treatment facilities which are already overburdened (Interview with GIZ, 03-10-2012). As an official climate change project funded by the German federal government the project also targets to contribute to mitigate global warming by reducing GHG emissions. According to the calculations in the feasibility study the waste-to-energy plant will save a total 4,700 tons of CO_2-equivalent emissions per year (Hamburg Wasser Kompetenznetzwerk, 2010, IV).

Impact beyond the Partnership Project

So far, the wider impact of the project is still limited as the waste-to-energy plant had yet to be completed at the time of this study. Once the plant is constructed and commissioned the project partners will start promoting the project in order to disseminate the technology. As stated above, Hamburg Wasser plans to bring German city officials to Nashik and teach them about the benefits of the co-fermentation process. The GIZ has not yet started publicising the project extensively in India as the plant first needs to prove to be functional in practise in the Indian context as a GIZ employee states. (Interview with GIZ, 27-11-2013). He adds that even though the GIZ has not started an official communications campaign, the project has already attracted a lot of interest in India (ibid.). He also explains that there is much interest in waste-to-energy technologies in India, despite a number of failed pilot projects giving waste-to-energy solutions for wastewater a rather bad reputation in the country (ibid.). However, the Indian government is putting cities under pressure to experiment with waste-to-energy technologies in order to generate energy. Furthermore, several Indian states promote the waste-to-energy approach as a solution for improved septic management which is increasingly identified and acknowledged as a crucial challenge in India. The GIZ employee adds that their partner university, the Goa Birla Institute of Technology and Science, has also started to promote the technology locally (ibid.).

Another GIZ employee involved states that officials from many other cities have already voiced their interest in the waste-to-energy plant. He points out that many visitors regularly come to Nashik to learn more about the existing MSW plant and in the future the GIZ plans to introduce them to the waste-to-energy technology, as well. The GIZ will also conduct dissemination workshops as soon as construction work has started (Interview with GIZ, 03-10-2013).

(Perceived) Partnership Mutuality

Like the tramway partnership between Pune and Bremen, the waste-to-energy initiative scores highly with regard to mutual project interests. Both Nashik and Hamburg Wasser are clearly driven by self-interest and expect to benefit from the project.

One of Hamburg Wasser's major motivations for joining the project was to demonstrate the Hamburg Water Cycle®'s practical and political feasibility. The Hamburg Wasser consultants emphasise that Hamburg Wasser decided to take part in the project as Nashik offered a more favourable local institutional arrangement for a combined treatment of solid waste and wastewater than Hamburg (Interview with Hamburg Wasser, 01-03-2013).

The GIZ project manager confirmed that the Nashik city administration is responsible for both septic tank management and solid waste collection, as Indian cities are constitutionally responsible for waste collection and usually also for wastewater, while in Germany they are usually managed by separate departments (Interview with GIZ, 27-11-2013). By proving the feasibility of the technology in India, Hamburg Wasser aims to convince German decision makers to adopt the HWC® technology and to learn from Nashik how to achieve a more integrated water and waste management policy. A Hamburg Wasser consultant highlighted that, similar as the situation faced by their Indian counterparts, it is very difficult to convince German city managers of the value of a technological concept if it has not yet been tested in practice. He noted that he intends to use the waste-to-energy plant in Nashik as a demonstration facility once it is completed. He explained that if Hamburg Wasser can show that an innovative sanitation approach works in the challenging conditions of India this may help convince administrations and decision makers in German cities to adopt such a new approach (Interview with Hamburg Wasser, 01-03-2013).

The Superintendent Engineer of Nashik MC was aware that Hamburg Wasser is pursuing its own self-interests in its support for the project; he referred to the legal barriers that prevent the introduction of such an experimental W2E plant in Germany (Interview with Nashik MC, 30-09-2013).

All interview partners from Nashik stated that the waste-to-energy plant will benefit their city greatly. When asked about Nashik's cooperation with GIZ and Hamburg Wasser, a Nashik MC employee highlighted knowledge and financial grants as their most important gains (Interview with Nashik MC, 05-12-2012).

Post-project Sustainability of the City Partnership

Despite the successful collaboration between the German and Indian project partners, so far no further cooperation has been planned between Nashik and Hamburg Wasser. A GIZ employee explains that the main priority is the successful completion of the waste-to-energy project with post-project collaboration currently not on the agenda (Interview with GIZ, 27-11-2013). As the GIZ has moderated and controlled all interaction between the Nashik MC and Hamburg Wasser in the waste-to-energy project, there has been very little direct communication between the protagonists and personal relationships have not developed. As a result, it seems unlikely that the partnership will continue beyond this single project. A Hamburg Wasser consultant points out that India is not one of Hamburg Wasser's primary focus countries. While he is personally convinced that it may be important to continue engaging with India in the development of sanitation technology, India is not a priority region for the utility (Interview with Hamburg Wasser, 01-03-2013).

5.3.3 Partnership Project W2E: Explanatory Factors

5.3.3.1 Knowledge Exchange Strategy

The waste-to-energy partnership project between the GIZ, Hamburg Wasser and Nashik Municipal Corporation achieves the highest total score of all four case studies for the index 'knowledge exchange strategy'. The collaboration fulfils or largely fulfils four of the seven indicators (prior program and context evaluation, involvement of external moderator, utilisation of policy windows and incorporation of intrinsic interests) and it partially fulfils the three remaining indicators of incremental and systematic transfer, continuity of interaction and policy windows (see table 13).

Indicator: Knowledge Exchange Strategy	Project Evaluation: W2E (tentative)		
	Nashik	Hamburg	Total
Prior Program Evaluation	/	/	++
Prior Context Evaluation	/	/	++
Systematic Cooperation	/	/	+-
Continuity of Interaction	/	/	+-
External Moderator/Facilitator	/	/	++
Policy Window	/	/	+-
Intrinsic Interests	++	++	++

Table 13: Partnership Project Waste-to-Energy: Findings from Index Knowledge Exchange Strategy

Prior Program and Context Evaluation

In the preparatory and planning phase the partnership actors conducted a comprehensive analysis of the technology and the local context conditions for building the waste-to-energy plant in Nashik.

A Hamburg Wasser consultant reports that the GIZ deliberately involved Hamburg Wasser due to the utility's conceptual experience of waste-to-energy technology in the form of the Hamburg Water Cycle (Interview with Hamburg Wasser, 01-03-2013). He adds that prior to the project in Nashik Hamburg Wasser had tested the HWC® in a small application in a local fun park, the Erlebnispark Karlshöhe, and, simultaneously to the Nashik project, Hamburg Wasser implemented the HWC® in a larger pilot facility in Hamburg Jenfeld Au. In addition, Hamburg Wasser also drew experiences from the running of a plant featuring a similar technology that Hamburg Wasser has been operating in Lübeck Flintenbreite for about ten years. The Hamburg Wasser consultant explains that the HWC® was designed to be transferrable to foreign countries, as well (ibid.). A GIZ employee involved confirms that the GIZ engaged Hamburg Wasser for the project in Nashik due to the utility's practical experience with waste-to-energy plants (Interview with GIZ, 27-11-2013).

In the project's feasibility study the Hamburg Wasser consultants adapted the HWC® to the conditions and challenges of the urban Indian context and suggested the co-fermentation of organic kitchen waste and black water from municipal toilets. Both waste materials are available in abundance in Nashik, and are overburdening the existing local storage and treatment facilities. A GIZ employee involved confirms that this form of co-fermentation is a novel approach which Hamburg Wasser developed considering the Indian context (Interview with GIZ, 27-11-2013).

He adds that the GIZ also commissioned the pre-feasibility and the feasibility studies to analyse the economic framework conditions in India and explore possible business models for the waste-to-energy plant in Nashik (ibid.).

The feasibility study also provides an overview of the historic development of centralised waste and wastewater treatment systems in European cities and their benefits and shortcomings. The authors explain that centralised treatment is widespread in Europe and most sewage systems are operating well. But they point out that this approach is neither affordable nor suitable in the context of most developing cities, as centralised waste and wastewater treatment is highly energy and water intensive, it lacks nutrient recovery, requires high capital investment and the combined sewage and rainwater pipes can lead to overflow and pollution in the event of heavy rainfall. They present the HWC® as a technological approach that addresses all of these shortcomings and provides an efficient and flexible means to treat waste and wastewater in different local contexts (Hamburg Wasser Kompetenznetzwerk, 2010, appendix, XVI-XXI).

The Hamburg Wasser consultants introduced the mayor and commissioner from Nashik and other local officials to the HWC® concept at a workshop on October 19, 2010, where they presented the feasibility study and Hamburg Wasser's experiences with the waste-to-energy technology. In the following year the GIZ facilitated a visit by Nashik MC's Superintending Engineer to Hamburg where he joined an environmental law conference and visited Hamburg Wasser. The Hamburg Wasser consultants showed Nashik MC representative several waste facilities operated by Hamburg Wasser in Hamburg, which they believe helped a great deal in eventually securing Nashik's approval for Hamburg Wasser to join the project as a consultant (Interview with Hamburg Wasser, 01-03-2013).

An evaluation of the technology and the local context conditions was further specified in the project DPR which the Indian consultant Paradigm developed in consultation with Hamburg Wasser and submitted to the GIZ on Sept. 13, 2011. The report provides concrete guidelines for setting up the required waste collection infrastructure, for the construction and operation of the plant and for its environmental and legal compliance. It also further specifies the business model and contains a detailed cost overview (Paradigm Environmental Strategies Pvt. Ltd., 2011).

Systematic Cooperation, Continuity of Interaction and External Moderator/Facilitator

This section assesses the knowledge transfer process with a particular focus on whether the project has followed a systematic knowledge exchange approach, if there has been continuity in the direct exchange and what role the external facilitator, the GIZ, played.

Until the project was delayed by the missing approvals from the Nashik city council the cooperation between Hamburg and Nashik was largely systematic and incremental. Initially, the GIZ identified the demand and the potential for a waste-to-energy plant in the GIZ's partner city Nashik. The GIZ then successfully applied for funding at the German government's International Climate Initiative and conducted a pre-feasibility study analysing the general conditions for waste-to-energy plants in India. For a more in-depth analysis of the HWC® technology and its adaptation to the Indian and Nashik context, the GIZ commissioned the German consultant, Hamburg Wasser, to prepare a feasibility study and the Indian consultancy, Paradigm, to execute a detailed project report. After the submission of both studies the project implementation was delayed by the local council's approval process, which meant that at the time of completion of the study the project had not yet been finalised. As a result, the indicator of "Systematic, Incremental Transfer" is tentatively assessed to have been partially fulfilled.

The other indicator that has only been partly met measures the continuity of personal exchange between the protagonists from the partner cities. The GIZ facilitated several exchange visits between representatives of Nashik MC and Hamburg Wasser. The Hamburg Wasser employees conducted two visits to Nashik as part of the waste-to-energy project; in April 2010 a Hamburg Wasser consultant collected the data required to prepare the feasibility study and in October 2010 the Hamburg Wasser consultants presented the results of the study at a GIZ workshop on waste management in Nashik. In September 2011, the Nashik MC Superintending Engineer joined an international environmental law conference in Hamburg and was introduced to Hamburg Wasser's work and urban sanitation (Interview with Hamburg Wasser, 01-03-2013). The commissioner of Nashik, Sanjay Khandare, also visited Hamburg Wasser as part of the GIZ's technical learning tour to Hamburg and Berlin on sustainable urban sanitation between September 23 and October 3, 2012 held for around 15 Indian officials (Adelphi Consult, 2012). A GIZ employee states that once the construction of the waste-to-energy plant begins the GIZ plans additional exchange visits, especially for the project monitoring in Nashik which will be conducted by Consul Aqua Hamburg (Interview with GIZ, 05-12-2012).

This interviewee acknowledges that for in-depth learning more exchange would be required between the Nashik and Hamburg project partners. He points out that continual exchange among the protagonists would be required to maintain the partnership's momentum and that ideally Nashik MC staff should conduct long-term visits of two or three months to Hamburg so that they could learn more about the technology and the contextual conditions. He regrets that this could however not be realised in the waste-to-energy project partnership as the Nashik city administration lacked the financial resources to send its employees abroad (ibid.).

All personal exchanges between Nashik MC and Hamburg Wasser staff were enabled by the GIZ and BMU funding. At the same time the strong leadership role played by the GIZ in the project prevented more direct communication between the German and Indian protagonists. The GIZ was keen to keep control of all project interaction due to its responsibility for the project's outcome with regards to the funder, the BMU (Interview with GIZ, 27-11-2013). Thus, Nashik MC and Hamburg Wasser employees were only able to interact directly during the exchange visits to Nashik and Hamburg. These visits were organised by the GIZ who also moderated all meetings and even helped prepare the presentations (Interview with Hamburg Wasser, 01-03-2013). A Nashik MC employee involved explains that apart from the visits, he has not communicated with his partners from Hamburg (Interview with Nashik MC, 30-09-2013). A Hamburg Wasser consultant adds that there have been no direct official contacts between Hamburg Wasser and Nashik outside the GIZ framework (Interview with Hamburg Wasser, 01-03-2013). A GIZ employee confirms that all communication between the partners goes via the GIZ, explaining that the Hamburg Wasser consultants cannot always come to Nashik to solve practical challenges. She adds that this is the reason why she has been located in Nashik to coordinate and speed up the implementation process of the project (Interview with GIZ, 30-09-2013).

Unlike the Pune-Bremen partnership the collaboration between Nashik and Hamburg has been exclusively driven by the GIZ as external facilitator of the partnership. The GIZ was founded in 2011 as a merger between the three former state development organisations Gesellschaft für Technische Zusammenarbeit (GTZ), InWEnt - Capacity Building International and the Deutscher Entwicklungsdienst (DED). It is operating in 130 countries worldwide. As of December 2012, the GIZ had more than 16,000 employees and a business volume of around 2.1 billion

Euros (GIZ, 2013, 1). In India the GIZ and its predecessor organisations have implemented development cooperation projects for over 60 years. The organisation funds projects totalling around 25 million Euros per annum[49].

The waste-to-energy project in Nashik has been conducted under the framework of the GIZ's Indo-German Environment Partnership (IGEP). This programme builds on the former Advisory Services in Environment Management (ASEM) programme and focuses on fostering bilateral collaboration in the areas of environmental and climate protection as well as clean urban infrastructure development. In the IGEP sub-section Sustainable Urban Habitat (SUH) the GIZ supports Nashik and six other selected Indian cities in mainstreaming climate protection through municipal solid waste management (Mutz, 2012). In addition to the waste-to-energy project the GIZ has also collaborated with the Nashik Municipal Corporation in sewage, hydraulic modelling and solid waste management projects (Interview with GIZ, 30-09-2013 (2)).

In the waste-to-energy cooperation the GIZ functions as the project leader and coordinator. Hamburg Wasser has supported the project as a technical consultant without being involved in financial and political matters. A Hamburg Wasser consultant states that Hamburg Wasser and the GIZ developed the project idea together, but he emphasises that the GIZ clearly took the lead in the partnership project (Interview with Hamburg Wasser, 01-03-2013). He adds that the partnership was fully dependent on the GIZ, pointing out that Hamburg Wasser alone would have neither the capacity nor the political contacts and intercultural competencies to establish such a transnational partnership (ibid.).

A GIZ employee confirms that German city officials would find it very difficult to access decision makers in Indian cities and initiate collaboration without the involvement of mediators such as the GIZ. He explains that the GIZ considers itself as an international facilitator, also helping German and Indian officials with practical issues such as translating documents, getting visa and preparing the visitors for climatic, cultural and working differences in the partner countries' local contexts (Interview with GIZ, 03-10-2013). Another GIZ employee states that the GIZ takes over the role as intermediary and local support agency in Nashik, as the German consultants cannot be around all the time and the Nashik officials are overloaded with work. She concludes that a supporting third party such as the GIZ is required to initiate and push projects. When asked whether she thinks that a transnational urban collaboration such as the Nashik project could be established without a facilitator such as the GIZ, she points out that such cooperation may be possible, but more difficult to set up. According to her experiences, transnational

---

[49] https://www.giz.de/en/worldwide/368.html (30-06-2014)

urban cooperation requires some kind of facilitator that provides the required insights into local conditions, capacities and problems plus helps receiving state and central government approvals (Interview with GIZ, 30-09-2013).

Policy Windows and Intrinsic Interests

The political conditions for the waste-to-energy project have been mixed and a policy window of opportunity only opened when the local council in Nashik eventually approved the project in 2013, about two years behind the initial schedule.

The problem stream has been in favour of the project, as the demand for improved waste management and energy production is generally high in Indian cities. Due to increasing urbanization, solid waste and sewage streams are growing by the year in most cities and at the same time, many cities suffer from regular power cuts as the energy infrastructure is fragile and overburdened. The GIZ and Hamburg Wasser identified Nashik as having a particular need for improved organic waste management. The existing composting plant is often shut down because of power cuts and therefore only treats about five percent of the city's organic waste, leading to high methane emissions (Interview with Hamburg Wasser, 01-03-2013).

Also the policy stream has been rather favourable as the waste-to-energy technology appears to provide an adequate solution to local needs. The GIZ project coordinator highlights that a key target of the waste-to-energy project is to demonstrate the feasibility of larger-scale but decentralised waste management in the Indian urban context. He explains that India is desperately looking for alternative solutions to Western-style centralised waste treatment which is often not affordable in Indian cities, despite the large grants provided for it by the international donors such as the World Bank, the Asian Development Bank (ADB) or the Kreditanstalt für Wiederaufbau (KfW) (Interview with GIZ, 27-11-2013). The HWC© concept for Nashik has been designed as an economical and decentralised approach to addressing the challenges of the waste treatment situation in Nashik. The project partners expect that the waste-to-energy plant will improve the local sanitation conditions and reduce the pressure of the existing composting plant (Interview with Hamburg Wasser, 01-03-2013).

The political stream has however not continuously been in favour of the partnership project. The GIZ could utilise the positive political conditions for the project from the side of Germany. The GIZ identified the International Climate Initiative funding framework as an adequate scheme for the waste-to-energy project and successfully applied for the grant. Moreover, the German partner, Hamburg Wasser, was interested in demonstrating the feasibility of its HWC® technology in the Indian city context. The utility was therefore willing to join the project without expecting immediate profits from it (Interview with Hamburg Wasser, 01-03-2013). The political conditions for urban sanitation programmes are also rather positive at the national level in India. The Indian government has set up a large-scale funding scheme for urban infrastructure projects, called the Jawaharlal Nehru Urban Renewal Mission (JNNURM). Under this framework the GIZ has supported Nashik's city administration in developing a comprehensive City Sanitation Plan (Interview with GIZ, 27-11-2013). The cooperation between the GIZ, the local commissioner and Nashik's city administration has generally worked well. The commissioner himself even got engaged in the project, promoting its benefits in a presentation at the local council (Interview with Nashik MC, 30-09-2013). It was only the local council that was more sceptical towards the project and therefore hesitated in giving its approval (Interview with GIZ, 30-09-2013). As the project required the consent of Nashik city council to initiate the tendering process, the lack of approval held up implementation considerably and led to delays in the project schedule.

A GIZ employee explains that after the local council's initial scepticism was overcome, most people in Nashik were convinced of the project, considering it a suitable and convenient way to degrade the biodegradable waste in a financially profitable manner (Interview with GIZ, 30-09-2013). The partners directly involved in the project, Hamburg Wasser and the Nashik MC, have pursued complementary interests. For Nashik MC, the waste-to-energy plant offers a low-cost and eco-friendly way to improve local sanitation and waste management conditions. The GIZ project coordinator highlights that Nashik, like many other Indian cities, is urgently searching for alternative models for centralised waste and wastewater treatment (Interview with GIZ, 27-11-2013).

A Hamburg Wasser consultant confirms that growing urbanization in India has put cities under pressure to acknowledge waste and wastewater treatment as a key priority. He adds that communication in Nashik regarding the waste-to-energy project has mainly focused on its co-benefits in the areas of hygienic sanitation, waste management, energy production and financial profit, rather than its positive impact on the climate (Interview with Hamburg Wasser, 01-03-2013). A GIZ employee confirms that the GIZ has promoted both the project's benefit to the climate as well as other co-benefits, as climate protection alone would not suffice in garnering public approval in the Indian context (Interview with GIZ, 27-11-2013).

A Nashik MC employee highlights that Nashik MC and the GIZ provided the local council with a complete picture of the project, including the low-carbon aspect, but he admits that climate change has remained a topic of little interest and priority for council members (Interview with Nashik MC, 30-09-2013).

On the German side, the project has a stronger focus on climate protection. It is funded via the German government's ICI program and listed on the ICI homepage. The reduction of GHG emissions and the generation of clean energy are also highlighted as major project objectives in an ICI leaflet describing the waste-to-energy project in Nashik (Dube, 2012).

Hamburg Wasser is not only interested in the project from a climate mitigation perspective, but also from a business point of view. A key driver for Hamburg Wasser's involvement in the initiative is to test and demonstrate its HWC® technological approach in a Global South city context. A Hamburg Wasser consultant explains that part of Hamburg Wasser's motivation to develop the HWC® was to transfer the technology abroad and thereby strengthen the utility's international presence. Furthermore, the specific co-fermentation design of the waste-to-energy project in Nashik cannot be easily tested in Germany due to administrative constraints. Hamburg Wasser intends to bring officials from Hamburg and other German cities to Nashik once the plant is commissioned to demonstrate to them that a co-fermentation of organic waste and black water is technically feasible and profitable, thereby building the foundations for potential new business opportunities Hamburg Wasser in Germany (Interview with Hamburg Wasser, 01-03-2013).

5.3.3.2 Linkages between the Partnership Project and State Institutions

The partnership between the GIZ, Hamburg Wasser and Nashik Municipal Corporation scores the highest of all four case studies in the index that measures the second independent variable, the institutional linkages between the project and the state system. It fully (++) or largely (++-) meets four of the six indicators, while the remaining indicators, approval by local state institutions and commitment of local state leaders, are partially fulfilled (see table 14).

Indicators: State Institutionalisation	Evaluation: Waste-to-Energy (tentative)		
	Nashik	Hamburg	Total
Formal Approval by Local State Institutions	++	--	+-
Formal Approval by (Sub)National State Institutions	++	++	++
Public Human Resources	+-	++	++-
Public Financial Resources	+-	++	++-
Public Coordinator	++	++	++
Commitment of Local State Leaders	+-	+-	+-

Table 14: Partnership Project Waste-to-Energy: Findings from Index State Institutionalisation

Formal Approval by Local and (Sub)National State Institutions

From the outset the waste-to-energy project was formally embedded into state institutions in both Germany and India.

At the local level, the project is part of the broader collaboration between Nashik Municipal Corporation, the local state body, and the GIZ to develop a city-wide sanitation policy. The sanitation policy aims to "Capture and treat all black and grey water to prescribed standards and incorporate recycling and re-use to conserve fresh water resources" (Nashik Municipal Corporation, 2011, 41). The major short-term targets of the policy include providing the entirety of Nashik with a piped sewerage network (up from 80% in 2011) and improving wastewater collection and treatment processes within three years. In the medium-term (within five years), Nashik aims to provide 100% of its households with individual sewerage connections (up from 90% in 2011) and reuse or recycle 20% of the city's wastewater (up from 0% in 2011) (ibid.). The waste-to-energy project has been designed to be implemented in a public-private partnership model, featuring Nashik MC as the main local project partner. In the planning phase, the project required and eventually received approval by Nashik's City Council and the local Standing Committee. During the implementation phase Nashik MC has been in charge of finding a private contractor through a tendering process. Furthermore,

Nashik MC is responsible for the construction of the pilot plant in cooperation with the contractor and the GIZ and for the provision of the land for the facility, plus for guaranteeing the waste collection process once the plant is in operation (Memorandum of Understanding between GTZ-ASEM and Nashik Municipal Corporation, 2010).

Aside from Nashik the waste-to-energy project does not require formal approval by the local city authorities in Hamburg, as Hamburg Wasser serves as a paid technology provider and project management consultant in the partnership and does not have formal responsibility for the project's implementation. A Hamburg Wasser consultant considers the waste-to-energy collaboration in Nashik primarily as a GIZ project, and he explains that the initiative has so far not officially been listed as part of Hamburg's climate policy activities. He adds that he has not yet promoted the project widely in Hamburg, but he would however do so if it turns out to be successful (Interview with Hamburg Wasser, 01-03-2013).

More than any of the other three cases in this study the waste-to-energy partnership has also been embedded in central and state government institutions. The collaboration was initiated and coordinated by the German federal government's development association, the GIZ (see more details in the section on "State Leadership" below). Furthermore, the German Federal Ministry for the Environment lists the project under its International Climate Initiative. As of February 2016, the waste-to-energy partnership is one of a total 26 projects in India that the ICI program has approved.[50] The GIZ project coordinator explains that the GIZ must report on the project's progress to the BMU once a year, and that a prerequisite for the ICI funding was that the project is executed under German legislation, however with the Court of Arbitration probably in Nashik (Interview with GIZ, 27-11-2013).

The waste-to-energy project is also closely linked to the Government of India. The GIZ project coordinator states that the GIZ has signed an implementation agreement with the Indian Ministry of Environment and Forest. The approval required from state institutions in both India and Germany poses challenges to the GIZ, particularly since the project implementation has been delayed and the GIZ needs to renew the implementation agreement with the Indian Ministry of Environment and Forest and obtain other administrative approvals (ibid.).

---

50   http://www.international-climate-initiative.com/en/nc/projects/projects/ (19-02-2016)

Public Human and Financial Resources

The waste-to-energy partnership project has been largely driven by state resources. Particularly in the planning and preparatory phase, the initiative was mainly executed by state employees and sponsored by state funds.

The GIZ personnel involved in the project are employed by the German government. The two Hamburg Wasser employees are also public employees. Hamburg Wasser was founded in 2006 as Germany's biggest public drinking and wastewater utility[51] and it is publicly financed by licence fees paid by Hamburg citizens (Interview with Hamburg Wasser, 01-03-2013). With the exception of the private consultancy Paradigm the project was also carried out by state actors in India, by GIZ employees, plus members of Nashik Municipal Corporation.

Funding for the project's planning phase, including exchange visits, the consultancy work by Hamburg Wasser and Paradigm, and the GIZ's coordination work were covered by the German government's ICI grant which totalled 2,036,382 Euros.[52] For the construction and operation of the plant, the GIZ decided to rely to a greater extent on private actors and set up a public-private partnership with a private contractor in order to relieve the overworked Nashik MC staff (Interview with GIZ, 30-09-2013 (2)). The contractor will also cover the funding gap for the construction of the plant and take over the financial responsibility for its operation and maintenance. The majority of the capital costs for the construction of the plant will however be covered by the ICI grant (around 0.8 million Euros). Also the operation will be mainly financed via state funds; Nashik MC will pay the contractor tipping fees for the waste collection and the state of Maharashtra provides feed-in-tariffs for surplus electricity produced.

With regard to the diffusion of the technology to other Indian cities the GIZ project coordinator emphasises that according to his calculations such a waste-to-energy plant could also be built and run without external subsidies. He explains that the current business plan is based on conservative cost estimations and leaves room to adapt it to other, potentially worse conditions, than those in Nashik (Interview with GIZ, 27-11-2013).

Public Coordinator and Commitment of Local State Leaders

State actor leadership can be assessed as mixed in the waste-to-energy partnership initiative. While the project has been coordinated and driven by the state institution, the GIZ, not all local state actors have continuously supported the project.

---

51 http://www.hamburgwasser.de/konzern.html (02-07-2014)
52 http://www.international-climate-initiative.com/en/about-the-iki/iki-funding-instrument/ (24-06-2014)

5.3 Case Study Nashik – Hamburg: Reducing Emissions through Waste-to-Energy

The project has been led and coordinated by the state institution GIZ. The GIZ is formally run as a private company but it is fully owned by the German government and it mainly implements development cooperation projects for the German Federal Ministry for Economic Cooperation and Development (BMZ). In agreement with the German government the GIZ can also work for other German ministries, other countries' governments, international donors or private companies (Stockmann, Menzel and Nuscheler, 2010, 432-436).[53] According to a Hamburg Wasser employee, the leadership role played by the GIZ has been crucial for the project's progress, as thanks to the GIZ the project members have had good access to key political decision makers in Germany and India (Interview with Hamburg Wasser, 01-03-2013).

While the GIZ has taken full ownership of the project, the commitment of the local state leaders was not always evident. This was particularly apparent in Nashik where the commissioner and the city administration have continuously supported the project, but the mayor and local council have at times been less proactive. The interest and willingness of Nashik's city administration to realise the waste-to-energy project was one of the reasons the GIZ selected Nashik for the project, a GIZ employee points out. He adds that in particular the head of Nashik MC, Commissioner Khandare, has shown strong support for the project. The GIZ employee puts this interest down to Khandare having an engineering background and a good understanding of waste-to-energy technology (Interview with GIZ, 07-12-2012). A Nashik MC employee confirms that Khandare has proactively supported the project, not only due to his engineering background but also because of his experience working in environmental policy making as the member secretary of the State Pollution Control Board (Interview with Nashik MC, 30-09-2013). Khandare publicly promoted the project and together with Nashik MC's Superintending Engineer gave a presentation to the local city council explaining the functioning and benefits of the waste-to-energy approach, which according to a GIZ employee was instrumental in eventually securing the council's approval for the project (Interview with GIZ, 03-10-2013).

Gaining approval from the local council has however proven to be one of the key challenges faced by the project and it has been a major reason for the delay in the project's implementation. A GIZ employee explains that members of the local council and Standing Committee were lacking the technological knowledge to understand the waste-to-energy concept and that the GIZ needed to invest a lot of time to explain the concept to the local political decision makers (Interview with GIZ, 30-09-2013 (2)). The GIZ project coordinator adds that apart from lacking understanding some sceptical council members deferred political approval and the

---

53   https://www.giz.de/de/ueber_die_giz/1689.html (02-07-2014)

decision was further held up by several local elections and the replacements of council members (Interview with GIZ, 27-11-2013).

A Hamburg Wasser consultant reports that in Hamburg political support for conducting projects in India has also not always been high. While Hamburg's former CDU-led senate (2008-2010) wanted to improve Hamburg's relations with India, the Social Democratic senate (in power since 2011) has shown less interest in the subcontinent. But as Hamburg's local state leaders have not been directly involved in the partnership project this differing level of interest has not had a major impact on the project's outcome (Interview with Hamburg Wasser, 01-03-2013).

5.3.3.3 Local Partnership Entrepreneurs

While the GIZ-facilitated partnership between Nashik and Hamburg in the waste-to-energy project scores highest of all four case studies in the indexes of 'Knowledge Exchange Strategy' and 'State Institutionalisation', the collaboration scores lowest in the index measuring the third explanatory variable, the existence of local partnership entrepreneurs. The total index score is actually zero as there have been no such individuals involved in this partnership project (see table 15).

Indicators: Partnership Entrepreneur	Evaluation: Waste-to-Energy (tentative)		
	Nashik	Hamburg	Total
Belief in Feasibility and Benefits	--	--	--
Investment of own Resources	--	--	--
Ability to Convince Stakeholders	--	--	--
Internal Policy Network	--	--	--
External Policy Network	--	--	--
Partner City Policy Network	--	--	--

Table 15: Partnership Project Waste-to-Energy: Findings from Index Partnership Entrepreneur

From the outset, the project was driven by the GIZ rather than by local individuals from Hamburg and Nashik. All the interview partners for this study referred to the

5.3 Case Study Nashik – Hamburg: Reducing Emissions through Waste-to-Energy 163

GIZ having sole responsibility for initiating and maintaining the partnership between the two cities. The project's protagonists from Hamburg and Nashik rely fully on the GIZ to facilitate knowledge exchange and up until now none of them has established any direct links with the respective partner city.

The Hamburg Wasser consultant is the only local individual who fulfils several key preconditions to serve as a partnership entrepreneur. He has a personal interest in the partnership project that goes beyond his function as a Hamburg Wasser representative and he is particularly convinced of the importance of engaging with the Indian market. He is also convinced of the feasibility of transferring the HWC® technology to India and he is generally open to innovative approaches. However, he also emphasises that he has not engaged in the project beyond Hamburg Wasser's defined responsibilities, due to his company's differing strategic priorities and the project's limited financial resources (Interview with Hamburg Wasser, 01-03-2013). He also points out that the ultimate responsibility for the project always lay with the GIZ (ibid.).

From the side of Nashik only the Nashik MC Superintending Engineer has been permanently involved in the waste-to-energy partnership project and the exchange with Hamburg Wasser. But beyond his official tasks he has not been personally engaged in the project, and he repeatedly referred to the GIZ being solely responsible for the project (Interview with Nashik MC, 30-09-2013).

When asked to name partnership entrepreneurs in the waste-to-energy project, a GIZ employee confirms that there are no such individuals involved and that the GIZ is fully in charge of driving the project (Interview with GIZ, 30-09-2013 (2))

5.3.3.4 Partnership Social Capital

Also with regard to the partnership social capital the waste-to-energy collaboration between Hamburg and Nashik scores lower than the Pune-Bremen partnership projects (see table 16). Nashik and Hamburg had never collaborated before joining the GIZ project and they could therefore not build on any prior partnership experience. Also more generally, many of the project protagonists from Hamburg and Nashik have limited experience in the field of international cooperation and they depended on the GIZ to overcome intercultural and communication barriers. Moreover the collaboration does not score highly with regard to stakeholder participation and inclusion. Only in the two remaining indicators measuring trust and equality among the partnership actors does the cooperation receive medium scores.

Indicators: Partnership Social Capital	Evaluation: Waste-to-Energy (tentative)		
	Nashik	Hamburg	Total
Prior Collective Action	/	/	--
Trust	+-	+-	+-
(Perceived) Equality	+-	+-	+-
Intercultural Experiences	--	+-	+--
Intercultural Communication	/	/	--
Involvement of Leadership Networks	--	--	--
Inclusion & Citizen Participation	+-	--	+--

Table 16: Partnership Project Waste-to-Energy: Findings from Index Partnership Social Capital

Prior Collective Action, Trust and (Perceived) Equality

Unlike the DEWATS and tramway initiatives between Pune and Bremen that benefited from the experiences and personal relations from prior partnership projects, the protagonists from Nashik and Hamburg had never collaborated before. There are also no other official relations between the cities of Hamburg and Nashik outside of this collaboration (Interview with Hamburg Wasser, 01-03-2013).

Both Nashik Municipal Corporation and Hamburg Wasser have cooperated with the GIZ on other projects. Nashik is one of the GIZ's focus cities in India and the GIZ has provided consultation services to Nashik on city-wide sanitation planning, solid waste management projects and non-revenue water management (Interview with GIZ, 05-12-2012). Hamburg Wasser has supported the GIZ in the development of an innovative septic tank and the GIZ and Hamburg Wasser jointly set up a feasibility study for a septic tank project in Cochin, Kerala (Interview with GIZ, 20-01-2013). Furthermore, the project coordinator knew Hamburg Wasser from her previous job in Hamburg's city administration (Interview with GIZ, 27-11-2013).

Due to the lack of prior project work and direct, personal relations between the project partners Hamburg Wasser and Nashik MC, there is little foundation for trust among these two partners. In the interviews, the Nashik MC representative and the two Hamburg Wasser project consultants emphasise that all their interaction has been moderated by the GIZ. The Nashik MC employee highlights that he has no direct personal relations with the Hamburg Wasser consultants (Interview with Nashik MC, 30-09-2013).

A Hamburg Wasser consultant recalls that it was only during the GIZ-facilitated exchange visits that Hamburg Wasser employees talked directly to Nashik Municipal Corporation officials. He states that the visit of Nashik MC's Superintending Engineer to Hamburg Wasser was crucial to gaining Nashik MC's approval for Hamburg Wasser joining the project as consultants.

But he also points out how difficult it is to sufficiently understand the cultural context and the local political processes of an Indian city such as Nashik without long-term engagement in the city itself (Interview with Hamburg Wasser, 01-03-2013).

The GIZ employees in the partnership project consider trusting relations a prerequisite for successful project cooperation. The two GIZ project coordinators visit Nashik on average every other month to conduct site visits and talk to local officials about the progress of the GIZ's sanitation and solid waste management projects. According to a GIZ employee this regular interaction is crucial to the development of trusting cooperation with the city officials (Interview with GIZ, 07-12-2012). Another GIZ employee confirms that the good relations between the GIZ and Nashik MC were established thanks to prior project experiences and the fact that the GIZ staffed the local Environmental Cell with employees who are from the region (Interview with GIZ, 30-09-2013 (2)). A GIZb project coordinator adds that the GIZ draws and depends on the trust and good reputation that the GIZ and German technology more generally has in India (Interview with GIZ, 27-11-2013).

The fact that the project protagonists from Hamburg and Nashik have rarely interacted directly without mediation by the GIZ has at the same time diminished the challenge of establishing equal relationships in the partnership. A Hamburg Wasser consultant is aware of the general problem of unequal relationships in North-South cooperation projects and that this is a sensitive issue in India. But he points out that this problem has hardly been evident in the waste-to-energy project as the GIZ has moderated the discussions between the project partners (Interview with Hamburg Wasser, 01-03-2013). A GIZ employee involved also believes that due to the fact that all project communication goes via the GIZ, issues of inequality have not emerged and hindered the project progress (Interview with GIZ, 30-09-2013 (2)).

Intercultural Competence and Communication

In comparison to the partners in the Pune-Bremen collaboration the project protagonists from Hamburg and Nashik are overall less experienced in conducting international cooperation projects. In particular the local officials from Nashik had little experience of working on international projects before joining the waste-to-energy partnership project. When asked about the main challenges faced by the project, a GIZ employee refers to Nashik officials' lack of exposure to international exchange (Interview with GIZ, 07-12-2012). A Nashik MC employee involved confirms that the waste-to-energy project is his first experience of international collaboration (Interview with Nashik MC, 30-09-2013).

The Hamburg Wasser representatives have some international experience and one Hamburg Wasser consultant explains that he was actually selected to take over the project as he has implemented projects in India before his time at Hamburg Wasser (Interview with Hamburg Wasser, 01-03-2013). In general Hamburg Wasser's international experiences are limited as most international projects are executed by their subsidiary, Consul Aqua Hamburg, who will also assume responsibility for monitoring the project implementation of the waste-to-energy partnership project.

As for the communication between the German and Indian project partners a Hamburg Wasser consultant points to differing approaches to giving presentations. He explains that he was not used to the Indian way of presenting and that he was also not always certain that the audience from Nashik fully understood his and his Hamburg Wasser colleagues' presentations which usually contained considerably less slides and text than those of the Indian partners (Interview with Hamburg Wasser, 01-03-2013).

The other Hamburg Wasser consultant adds that the project protagonists depended on the GIZ to overcome intercultural barriers, e.g. through giving both Hamburg and Nashik actors advice on setting up their presentations (ibid.). A GIZ employee elaborates that part of the GIZ's work in the project has been to translate presentations and documents into the local language Marathi and refers to the second and third-level officials in Nashik who sometimes lack the English language skills required for international collaboration (Interviews with GIZ, 07-12-2012 and 03-10-2013). The GIZ generally attaches high priority to language skills and prefers to staff its offices around India with local employees, another GIZ employee highlights and adds that the GIZ employees working at the GIZ Environmental Cell in Nashik are from the state of Maharashtra and know the local language Marathi (Interview with GIZ, 30-09-2013 (2)).

Involvement of Leadership Networks, Inclusion, and Citizen Participation

The collaboration between the GIZ, Hamburg Wasser and Nashik's city administration also does not score highly with regard to the two partnership social capital indicators of local stakeholder involvement and inclusion.

So far, local informal leadership networks from Nashik and Hamburg have not been involved in the project. A Nashik MC employee highlights that local NGOs have supported the preparation of the City Sanitation Plan in Nashik but they have not been actively involved in the waste-to-energy project as this was not institutionally required (Interview with Nashik MC, 30-09-2013). A GIZ employee confirms that apart from the four project partners, Nashik MC, the GIZ, Hamburg Wasser and Paradigm, no other local actors have participated in the partnership initiative (Interview with GIZ, 30-09-2013 (1)).

The authors of the project feasibility study point to the difficult challenge of finding the right balance in stakeholder involvement. On the one hand, they identified that the "multitude of responsibilities and stakeholders" have been key constraints in existing waste-to-energy plants in India. On the other hand, they emphasise that "the involvement of all stakeholders is crucial for the success of the project" (Hamburg Wasser Kompetenznetzwerk, 2010, 5).

The issue of the inclusion of minorities and underprivileged citizens has not been a primary focus of the partnership project, either. The slum population of Nashik has been one of the project's target groups; however local slum dwellers have not been directly involved in the project planning.

5.4 Case Study Nagur – Freiburg: The Local Renewables Model Community Network

5.4.1 Partnership Project LRMCN: Content and Process

The collaboration between Nagpur, India and Freiburg, Germany embodies a fourth, emerging form of urban partnership; the collaboration of cities in transnational municipal networks. Nagpur and Freiburg exchanged knowledge and ideas on low-carbon development as part of the Local Renewables Model Communities Network (LRMCN) which was run by the city network ICLEI-Local Governments for Sustainability (ICLEI) between November 2005 and June 2010.

The idea for the LRMCN project was developed at the International Conference for Renewable Energies in Bonn in June 2004 where ICLEI was in charge of organising the local government agenda. In cooperation with the German Federal

Ministry of Economic Cooperation and Development (BMZ) and the German government's development association, the GTZ, ICLEI established the Local Renewables Initiative as an operational framework to implement Local Agenda 21 processes with a focus on urban renewable energy development, energy efficiency and energy saving (ICLEI, 2010a). One of the Local Renewables Initiative's major projects has been the establishment of a series of Local Renewables Conferences in Freiburg (the conference has been hosted five times by the city of Freiburg since 2007 (Local Renewables Conference Website)). Another of ICLEI's main projects was to set up the LRMCN in 2005 to foster North-South collaboration in urban clean energy development. Via this network, ICLEI looked to develop model cities in countries of the Global South in cooperation with Northern partner cities, highlighting that both Northern and Southern cities can benefit from a clean energy transition (ICLEI, 2010a). The core aim of the initiative was to support and strengthen the role of local governments "as a driving force for innovation and investment in their communities." (ICLEI, 2007, 2)

The LRMCN involved three types of member cities: a) "Model Communities", that is, engaged cities from developing countries who formally committed to local clean energy development, b) experienced European "Resource Cities" which assisted the other network cities by offering their expertise in developing local clean energy strategies, and c) "Satellite Cities" which observed and learned from the LRMCN's activities, thereby enabling the wider diffusion of successful projects and policies (ICLEI, 2010b). ICLEI together with the project consultant, the GTZ, selected the two Indian ICLEI member cities, Nagpur and Bhubaneswar, as the first LRMCN Model Communities. ICLEI then affiliated these two cities with the Resource Cities of Freiburg, Bonn (Germany) and Vaxjö (Sweden); European ICLEI member cities that were already experienced in implementing clean energy strategies. With the financial help of the Global Opportunity Fund of the British government the LRMCN was further enlarged in 2005 to include the Brazilian cities of Betim and Porto Alegre (ICLEI, 2007). In 2008 the South Indian city of Coimbatore joined as a sixth Model Community and ICLEI also brought in Milan (Italy) and Malmö (Sweden) as two additional European Resource Cities. By 2009 the LRMCN thus encompassed five Model Communities and five Resource Cities complemented by eleven Indian and four Brazilian cities plus Cape Town (South Africa) who joined the network as Satellite Cities.

According to ICLEI employees working in the LRMCN project, Nagpur and Freiburg were among the most active LRMCN member cities and among the few Model and Resource Cities that conducted mutual exchange visits to both cities (Interviews with ICLEI, 08-08-2012 and 05-12-2013). These visits were facilitated by ICLEI in the early phase of the LRMCN project. In September 2006 Lokesh Chandra, the Municipal commissioner of Nagpur, joined an Indian delegation of

5.4 Case Study Nagur – Freiburg: The Local Renewables Model Community Network 169

city officials that travelled to Europe to visit Freiburg and Barcelona. An ICLEI employee explains that ICLEI designed this tour to introduce Indian city authorities to the concept of urban clean energy development. The goal was to motivate them to develop their own clean energy strategies by demonstrating the practical achievements of European frontrunner cities and facilitating peer-to-peer exchange between local decision makers. This ICLEI employee points out that Indian city officials in particular prefer to speak to counterparts of a similar hierarchical rank, so ICLEI put Chandra in touch with Freiburg's Mayor Dieter Salomon (Interview with ICLEI, 08-08-2012). ICLEI also organised a guided tour through Freiburg to show the Indian delegation Freiburg's progress in the fields of public transport, cycling infrastructure, energy-efficient buildings and urban planning (ibid.).

While the exchange visit to Freiburg mainly centred around mutual motivation and the political planning involved in local clean energy strategies, ICLEI fostered more practical exchange between technical experts from the Freiburg and Nagpur city administrations on the visit to Nagpur that followed. An administrative employee of the Energy Unit at the Environmental Protection Department in the City of Freiburg went to Nagpur for one week and joined an ICLEI workshop on local renewable energy development. He recalls that he and the Nagpur city administration staff responsible for energy management joined a workshop organised by ICLEI and they gave presentations on the state of the local clean energy development in both cities. The presentations were followed by more informal exchange of experiences, where the Nagpur staff asked the Freiburg representative practical questions on how to develop and implement clean energy measures and projects.

Amongst other things, the Freiburg representative provided advice to the workshop participants from Nagpur on the setting up of the Renewable Energy and Energy Efficiency Resource Centre, a technology demonstration centre and stakeholder meeting point which the Nagpur Municipal Corporation (Nagpur MC) had established in cooperation with ICLEI as part of the LRMCN. The Freiburg representative assessed this combination of presentations and informal talks on practical challenges as very insightful and fruitful and a good starting point for a potential longer-term exchange between Freiburg and Nagpur on clean energy development (Interview with Freiburg city administration, 07-08-2012).

In addition to these two visits ICLEI had indeed planned to facilitate more direct exchange between Freiburg and Nagpur as well as between the other LRMCN member cities. The ICLEI project employee states that ICLEI aimed to bring the LRMCN cities together at international conferences and workshops, such as the UNFCCC Conferences of Parties. This however turned out to be challenging as not all LRMCN members were able to join these meetings due to financial and time constraints (Interview with ICLEI, 31-10-2012)

This is exemplified by the Freiburg and Nagpur exchange, as representatives from both cities participated in several international ICLEI events during the time of the LRMCN project, but they never managed to join the same meetings.

Initially, ICLEI also intended to facilitate more mutual exchange visits between the LRMCN member cities, but over the course of the project ICLEI increasingly focused on utilizing its financial and time resources for realising concrete project implementation in the Model Communities rather than facilitating more direct exchange with the European Resource Cities. An ICLEI employee explains that one reason for concentrating on project work in the Indian cities was that the ICLEI project coordinators realised how difficult it is to enable direct learning between European and Indian cities, due to the very distinct communication cultures and legal competences of the respective cities with regard to clean energy development (Interview with ICLEI, 05-12-2013).

Rather than arranging direct personal exchanges ICLEI incorporated more indirect learning by drawing on the European cities' experiences in the drafting of project proposals for the Model Communities. The European Resource Cities served as a reference for the ICLEI staff in their development of locally tailored clean energy projects and policies in collaboration with Indian city administrations. The ICLEI employee highlights that the European cities' achievements in clean energy development were also used for the political marketing of the LRMCN projects in Nagpur and the other Model Communities. She explains that when the Nagpur city administration had ideas or even concrete proposals for clean energy projects it was very helpful to be able to refer to a Western partner city having successfully implemented a similar project in order to convince the local council and receive the required approvals (ibid.)

A member of the Nagpur MC staff who was in charge of executing the LRMCN program confirms that international exchange was a very significant source of inspiration, plus a psychological tool to convince local decision makers to back the projects. He points out that on several occasions references made to international cooperation projects have helped accelerate the decision-making process in Nagpur (Interview with Nagpur MC, 14-10-2013).

For the purpose of lesson-learning ICLEI compiled case studies on clean energy development in all LRMCN member cities. These case studies were circulated among decision makers in the network cities. Other than the direct exchange between the cities which mainly took place in the beginning of the project, the indirect learning via ICLEI documentation continued throughout the entire project period, as the ICLEI employee points out (Interview with ICLEI, 05-12-2013).

The project implementation in Nagpur was focused around realising the four key targets of the LRMCN for Model Communities: the conduct of energy and GHG emissions inventories, the establishment of clean energy information centres, the involvement of stakeholders, plus the introduction of city-level renewable energy and energy efficiency policies (ICLEI 2007).

The first target, the energy and emissions inventory, was already achieved in 2006. The inventory was established to provide a baseline for identifying areas of intervention as well as for the future monitoring of the LRMCN project's impact. The inventory assessed the city-wide energy consumption and GHG emissions both for the whole city as well as for the areas under direct control of the Nagpur MC. It revealed that in 2006 Nagpur emitted 1.76 million tons of CO_2 equivalents, resulting in per capita emissions of 0.88 tons per year. The emissions under control of the city administration – which were the main target of the LRMCN project – accounted for 4.8% of the city's total emissions. The major contributors to these corporate emissions were water and sewage systems (60%) and street lighting (33%). The results of the inventory were published in energy reports and updated annually over the course of the LRMCN project (ibid.).

To achieve their second project target ICLEI and the Nagpur MC in December 2006 established the Renewable Energy and Energy Efficiency Resource Centre (RC) as a competence centre and "catalyst" for clean energy development in Nagpur (ICLEI, 2010a, 9). The RC was located in the city administration's premises and the LRMCN project was coordinated by ICLEI and Nagpur MC employees. They prepared the energy and emission inventories as well as project proposals and supported Nagpur MC in the implementation and monitoring of these proposals. Furthermore, the RC also served as a clean energy demonstration and knowledge centre and a stakeholder meeting point. ICLEI considered local stakeholder involvement as key to the project's success in local clean energy development and this constituted the third project target. The LRMCN was thus designed to involve stakeholders both as consultants as well as project partners. (In this study stakeholder involvement is defined as one indicator for the explanatory variable of partnership social capital; please find more details in the section Stakeholder Involvement and Inclusion below.)

With the introduction of a clean energy policy the fourth main project target was also achieved. Nagpur was the first Indian city to formally adopt a city-level renewable energy and energy efficiency policy. The policy addresses all sectors which fall under the legal jurisdiction of the municipal corporation (including street lighting, water and waste management, building codes and public transport) and sets the target of reducing the municipal energy consumption by 20% and the overall city energy consumption by at least 3% by 2012 compared with 2005 levels (ICLEI, 2008). The policy was accompanied by a five-year action plan (2007-

2012) and annual activity plans that were based on the energy inventory and their performance was regularly reviewed according by Nagpur MC, the Resource Centre and local stakeholders (ICLEI, 2010a).

5.4.2 Partnership Project LRMCN: Outcomes

Despite the fact that direct exchange between Nagpur and Freiburg was limited to the early phase of the LRMCN cooperation, the partnership project delivered considerable results and achieves the highest total score in the evaluation of the outcomes of all the four partnership projects analysed.

Success Indicators	Project Evaluation: LRMCN		
	Nagpur	Freiburg	Total
Implementation Achievements	/	/	++
Budget and Schedule Performance	/	/	+-
Local Capacity Building	+-	--	+--
Benefits for Target Group	/	/	++
Impact	/	/	++
Mutuality	/	/	--
Post-project Sustainability	/	/	--

Table 17: Partnership Project LRMCN: Findings from success index

The collaboration scores highly with regard to implementation achievements, benefits for the target group and its wider impact. Plus, it achieves partial success in budget and schedule performance and in capacity building in Nagpur. The lack of more comprehensive cooperation between the two cities is however reflected in the project's low scores for capacity building in Freiburg, partnership mutuality and post-project sustainability (see table 17).

Implementation Achievements

The LRMCN was the only one of the four partnership initiatives analysed to realise implementation on both micro and macro levels. During the project, between 2005 and 2010, Nagpur MC and ICLEI set up a local information and resource centre, the partners prepared a city-wide energy and GHG emissions inventory and Nagpur MC formally committed to clean energy promotion. The fact that Nagpur was the first city in India ever to introduce a local clean energy policy can be regarded as a positive project achievement in itself (which led to Nagpur being selected as a model city in the Indian Solar Cities Program and thereby having a nationwide impact, see section Impact beyond the Partnership Project below). With reference to the inventory and city policy, Nagpur MC and ICLEI conducted a multiplicity of small-scale demonstration and awareness-raising projects, many of which focused on local solar energy promotion. The projects included a solar water heater system built in the municipal maternity hospital, the construction of a solar-powered power back-up at the Nagpur Resource Centre and the installation of energy efficient lighting at a municipal girls' school (ICLEI, 2010a).

In cooperation with Nagpur MC and local stakeholders the RC also conducted a total of nine awareness-raising projects in Nagpur, for example the Renewable Energy Day which was celebrated in Nagpur on August 20, 2007, including an exhibition and workshops on local renewable energy usage, an essay competition in which 5000 pupils participated and a Renewable Energy Run. Throughout 2007-2009, several renewable energy and energy efficiency trainings were provided to local stakeholders such as women's groups, schools and Nagpur MC staff. A third example was the mobile promotion of Nagpur's solar water heater funding scheme in 2010 (ICLEI 2010a).

In addition, two large-scale projects were realised as part of the LRMCN program. In cooperation with ICLEI and the Government of India's JNNURM, Nagpur MC undertook a water and energy efficiency audit at five water treatment plants and two pumping stations in 2005. Moreover a wastewater reuse system was installed at local thermal power plants in 2008 in collaboration with the JNNURM and the state power supplier, the Maharashtra Generation Company (Mahagenco) (ICLEI 2010a; ICLEI 2010c).

A member of the Nagpur MC staff states that despite these projects the energy city policy targets (20% reduction in the municipal energy consumption and 3% reduction in the overall city energy consumption) had not yet met been met by the time of the completion of the project in 2010. He explains that the targets were highly ambitious and that Nagpur required more time to achieve them. He still considers the project a success and highlights that the energy audits have made a particularly strong impact on the work of Nagpur's city administration as they enabled the Nagpur MC staff to understand what kind of energy savings are possible, especially in the areas of water and sewage treatment (Interview with Nagpur MC, 14-10-2013)

ICLEI also concludes that the project implementation in Nagpur was on the whole successful. The organisation considers all four indicators established at the beginning of the project as being fulfilled, namely: the development of a local clean energy strategy and the commitment to concrete targets in a local policy (indicator 1); the establishment of a local information centre (the RC) (indicator 2); project implementation in the areas of short-term, long-term and awareness-raising initiatives (indicator 3); and Nagpur's integration into national and international local government networks (indicator 4) (ICLEI 2010a).

With regard to the fourth indicator, national and international networking, ICLEI's final report on the LRMCN implementation in Nagpur lists a number of national and international conferences attended by Nagpur representatives, such as the United Nations Climate Conferences COP 14 in Poznan (December 2008) and COP 15 in Copenhagen (December 2009), the World Urban Forum in Nanjing (November 2008) and the South Asia Regional Meeting for the Climate Roadmap in Delhi (October 2009) (ibid.). An ICLEI employee states however that ICLEI decided to withdrew initial plans for more cooperation and exchange between Nagpur, Freiburg and the other European Resource Cities when ICLEI realised how difficult it was is to facilitate direct learning (see above).

Budget and Schedule Performance

In Nagpur, the majority of LRMNC activities were implemented within the project's scheduled timeframe of 2005 to 2010. The RC was opened in December 2006, the city-level renewable energy and energy efficiency policy was introduced in August 2007 and several pilot, large-scale and awareness-raising projects were realised before the project completion in June 2010 (see section entitled Partnership Project Local Renewables Model Communities Network: Content and Process above). An ICLEI employee states however that some of the local projects in Nagpur were delayed by several months and beyond the scheduled completion of the LRMCN due to local elections and changing local decision makers (ICLEI in email correspondence, 26-09-2014). Several energy efficiency and renewable energy projects that were developed in 2009 and 2010 had still not been completed at the point of data collection for this study. The ICLEI employee points out that Nagpur may be able to take them up again when the city receives additional resources as one of the Solar Cities Program's model communities (Interview with ICLEI, 05-12-2013).

5.4 Case Study Nagur – Freiburg: The Local Renewables Model Community Network   175

As for budget performance the LRMCN scores well, as the initial project funding by the BMZ was not exceeded. The GTZ agreed to issue a cost-neutral extension for implementing several projects that were still pending by the end of the project period (ICLEI in email correspondence, 26-09-2014).

Local Capacity Building

The partial achievements of the LRMCN project in the area of local capacity building are restricted to the city of Nagpur, as there was no capacity building in Freiburg.

In Nagpur, the LRMCN fostered institutional reform with the city by formally committing to clean energy development in the Renewable Energy and Energy Efficiency Policy. Nagpur MC employees also developed technical as well as intercultural skills through their participation in the LRMCN projects and the visits to the partner cities and international conferences (Interview with Nagpur MC, 14-10-2013). Moreover, a local Resource Centre was set up in the City Hall "to establish a competence Center for energy efficiency and renewable energies with qualified staff within the city government." (ICLEI, 2010a, 35) Throughout the LRMCN project period the RC was hosted by Nagpur MC and jointly run by the city administration and ICLEI staff (Interview with ICLEI, 05-12-2013). Furthermore, both Nagpur MC and ICLEI employees highlight that the energy and emissions inventories were crucial to establish a baseline and provide guidance where to focus the attention of the LRMCN projects (Interview with Nagpur MC, 14-10-2013 and ICLEI, 05-12-2013). Throughout the project period the inventory was annually updated by an ICLEI employee who ran the RC in Nagpur. The project coordinator of ICLEI however points out that the updates remained dependent on this ICLEI employee, and that Nagpur MC has not developed enough capacity to run the inventory by itself. This became apparent when the employee left the city after the project's completion and Nagpur MC staff found it difficult to keep the inventory updated. (Interview with ICLEI, 05-12-2013)

The ICLEI employee working at the RC also strongly supported Nagpur MC with the preparation and agenda setting of the LRMCN project proposals as the city administration lacked the capacity and time to assume sole responsibility for them. Similarly, the Resource Centre remained strongly dependent the ICLEI employee and when the LRMCN project was completed and the employee left, the RC became much less active (ibid.).

Freiburg did not build any capacity in the LRMCN program. The member of the Freiburg city administration who visited Nagpur points out that he enjoyed travelling to Nagpur. But the exchange as part of the visit was not enough for him to develop a comprehensive understanding of Nagpur and its energy strategy. He adds that he was hoping to visit Nagpur more often to build the ground for concrete project collaboration. But after conducting several failed attempts to continue the exchange with Nagpur he eventually stopped engaging in the LRMCN program. Other forms of capacity building have not taken place in Freiburg (Interview with Freiburg city administration, 07-08-2012).

Benefits for Target Groups

As the focus of the LRMCN program was primarily on the development of the Indian and Brazil Model Communities, the major target groups were also located in these cities. ICLEI defined the Model Communities' local governments as the most important target group of the LRMCN:

> The main goal of the Local Renewables Model Communities Network project (or the Local Renewables project) is to support and strengthen local governments (in India, these are called "municipal corporations") which promote the generation and supply of RE [renewable energy] and EE [energy efficiency] in the urban environment. The focus of the project is on the roles and responsibilities of municipal corporations as the driving force for innovation and investment in their communities. (ICLEI 2010a, 13)

The Nagpur MC employee responsible for the project points out that the Nagpur city administration has strongly benefited from the collaboration with ICLEI as part of the LRMCN. He highlights the example of the water and energy efficiency audit at Nagpur's local water power plants and ICLEI's advice on how to change the machinery that led to efficiency improvements of 50/60% to 90% at the plants. He also refers to the benefits of the demonstration projects at the municipal facilities and his personal exposure to international exchange, plus the motivation and inspiration that he and his colleagues from Nagpur MC gained from the collaboration with ICLEI in the LRMCN (Interview with Nagpur MC, 14-10-2013).

Project benefits for other LRMCN target groups such as local businesses and school children in Nagpur can also be identified. The RC measured that local companies engaged in renewable energy and energy efficiency services realised growth rates of 8-10% and in some cases more throughout the LRMCN project period: "the technology that reported the highest growth in sales was SWH [solar water heaters], while dealers of other RE and EE equipment also reported a growth in business of 8-10% over the past four years." (ICLEI 2010a, 37) School children were another key target group in the LRMCN. ICLEI and the Nagpur MC jointly developed a training program to educate pupils on renewable energy and energy efficiency which was carried out at a total of 25 local schools in

Nagpur. Moreover, the RC facilitated the creation of "energy conservation clubs" in 20 local schools where energy-related topics were discussed and pupils were trained in energy-conscious behaviour. Furthermore, a total of 20,000 students from Nagpur participated in a Renewable Energy Day on 22 August 2007 (ICLEI 2010a).

Impact beyond the Partnership Project

The LRMCN partnership project also scores highly with regard to its wider impact. Locally, Nagpur MC provided a separate fund of one million Indian Rupees to the RC and ICLEI to initiate additional renewable energy and energy efficiency projects beyond the LRMCN's completion year of 2010. One project supported by this fund was a proposal for energy-efficient lighting in Nagpur MC's sewage treatment plant which was designed based on the positive results of the LRMCN school lighting project (ICLEI 2010a).

At the national level in India, Nagpur has repeatedly served as a model city for other cities that engage in local clean energy development. According to the Government of India, several Indian cities such as Aurangabad and Thane have established water sector energy audits based on Nagpur's model project under the LRMCN. During the LRMCN Nagpur also shared its experiences in energy-efficient water management with the other network members and guided the third Indian LRMCN Model Community, Coimbatore, in preparing a tubewell energy audit (ICLEI 2010c).

The largest impact and most important legacy of the LRMCN were achieved when the network served as model for the Government of India's Solar Cities Program. In 2008 the Ministry of New and Renewable Energies (MNRE) became interested in the potential of cities to foster renewable energy development and decided to set up the Solar Cities Program, an initiative to promote solar energy at the local level. An ICLEI employee recalls that the MNRE approached ICLEI to support the design of the Solar Cities Program due to ICLEI's experiences with the implementation of the LRMCN in the Indian Model Communities (Interview with ICLEI, 31-10-2012). She explains that the Solar Cities Program adopted several structural components from the LRMCN implementation in the Indian Model Communities, such as the energy inventory, a policy and master plan based on the results of the inventory, the set-up of "Solar Cities Cells" (which were based on the RC) and the focus on awareness-raising projects in local communities (Interview with ICLEI, 05-12-2013).

The MNRE selected a total of 60 Indian cities to join Solar Cities Program and all received funding for preparing and implementing local solar energy strategies, among them the three Model Communities and most of the LRMCN's

Indian Satellite Cities. Nagpur was the first city to be accepted and in 2009 the city was even selected as one of two Model Solar Cities (ICLEI 2010a). According to an ICLEI employee this was to a large degree due to the initiatives and experiences Nagpur gained as part of the LRMCN project (Interview with ICLEI, 05-12-2013). A Nagpur MC employee highlights that as a Model Solar City Nagpur received additional funding totalling 90 million Rupees from the MRNE to develop solar energy locally, which was matched by another 90 million Rupees provided by Nagpur MC. He points out that the funds are being used to realise unfinished LRMCN projects and he is confident that they will help Nagpur to eventually achieve the targets set in the local renewables policy in 2007 (Interview with Nagpur MC, 14-10-2013).

(Perceived) Partnership Mutuality

While the LRMCN partnership initiative performs well with regard to implementation achievements in Nagpur and its wider impact in India, it scores less well in the area of partnership mutuality. The German partner city of Freiburg has not directly benefited from the exchange and not initiated any projects under the LRMCN. The member of the Freiburg administration who visited Nagpur states that he found it difficult to identify common ground for concrete project work in the cooperation with Nagpur. He highlights that it is generally important for Freiburg to be able to connect to partnership initiatives with its own ongoing projects to be able to justify the time and financial resources spent on city collaboration.

He points out that Freiburg with its longstanding experience in climate and clean energy action usually serves as the model city in many of its international collaboration projects. Freiburg has however started to shift its focus towards more mutuality in partnerships and he refers to the partnership with the French city of Besancon from which Freiburg recently adopted a program called "200 Familien für den Klimaschutz". He emphasises that even a frontrunner city such as Freiburg can benefit greatly from exchange with other cities. He identifies fundraising, lobbying and pragmatic administration as areas in which Freiburg can learn a lot from other cities (Interview with Freiburg city administration, 07-08-2012).

The ICLEI project coordinator confirms that on the exchange visit to Nagpur the Freiburg representative voiced the fact that Freiburg also wanted to learn from Nagpur (Interview with ICLEI, 08-08-2012). This mutual learning however failed as the direct exchange between the two cities stopped shortly after the exchange visits. Moreover, mutuality was not a key priority in the LRMCN's structural concept. Rather the LRMCN focused on developing renewable energy

5.4 Case Study Nagur – Freiburg: The Local Renewables Model Community Network 179

and energy-efficiency model cities in the Global South with the support of European Resource Cities which ICLEI defines as "Cities or towns that are ready to share their expertise and experience with other cities in the Network and beyond." (ICLEI 2010b, 13) Project development in the European cities was not a major focus of the LRMCN program.

Post-project Sustainability of the City Partnership

The partnership between Nagpur and Freiburg also receives low scores for its post-project sustainability. Similar to the exchange between Nashik and Hamburg which strongly depended on the project coordinator, the GIZ, all interaction between Nagpur and Freiburg in the LRMCN was facilitated by ICLEI. As described above, the direct exchange between Nagpur and Freiburg was limited to the early phase of the LRMCN and the two cities have not collaborated in any form since ICLEI completed the program in 2010. An ICLEI employee states that further cooperation between Nagpur and Freiburg would be desirable but that ICLEI depends on external funding for facilitating such exchange, which is currently not available. Therefore there is no concrete plan for further direct exchange between Freiburg and Nagpur via ICLEI (Interview with ICLEI, 05-12-2013).

5.4.3 Partnership Project LRMCN: Explanatory Factors

5.4.3.1 Knowledge Exchange Strategy

Despite its relatively successful outcome in Nagpur, the partnership between Freiburg and Nagpur in the LRMCN program scores comparatively low with regard to the existence of a knowledge exchange strategy, the first explanatory variable. The city partnership particularly lacked systematic and continuous direct exchange and scores well in only one indicator, the involvement of an external partnership facilitator, the city network ICLEI (see table 18).

Indicator: Knowledge Exchange Strategy	Project Evaluation: LRMCN		
	Nagpur	Freiburg	Total
Prior Program Evaluation	/	/	+-
Prior Context Evaluation	/	/	+-
Systematic Cooperation	/	/	--
Continuity of Interaction	/	/	--
External Moderator/Facilitator	/	/	++
Policy Window	/	/	+-
Intrinsic Interests	++	--	+-

Table 18: Partnership Project LRMCN: Findings from Index Knowledge Exchange Strategy

Prior Program and Context Evaluation

Only in the initial phase of the LRMCN program, did Nagpur and Freiburg officials conduct a mutual exchange visit to each other's' cities, during which the city representatives were introduced into each other's clean energy activities. These two visits were however not sufficient enough to enable a comprehensive and in-depth evaluation of the partner city's energy strategies and context conditions. This became particularly apparent when the Commissioner Chandra and a senior technical officer visited Freiburg. The ICLEI project coordinator remembers that she and her ICLEI colleagues organised a city tour to showcase the benefits of Freiburg's cross-sectoral climate action planning, which was a completely new approach for the Indian visitors who were not used to address energy, building, transport and urban planning approaches holistically (Interview with ICLEI, 08-08-2012).

The integrated manner of climate action planning across sectors was not only a novel perspective for the Nagpur representatives. They also did not fully understand all aspects and components of Freiburg's climate strategy and the ICLEI project coordinator realised that while the Indian visitors could connect with some of the initiatives such as public transport, other areas such as Freiburg's energy-efficient building projects remained obscure to them (ibid.)

To improve understanding of the different stages and approaches towards clean energy development, ICLEI conducted brief case studies of the LRMCN

5.4 Case Study Nagur – Freiburg: The Local Renewables Model Community Network   181

member cities highlighting areas of intervention and achievements. The case studies introduce Freiburg's cross-sectoral approach towards urban climate planning (ICLEI 2009) and highlight Nagpur's focus on energy-efficient water management (ICLEI 2010c). ICLEI distributed the case studies among the network members, "to serve as a means of information dissemination and also to serve as a knowledge resource for other cities" (ICLEI 2010b, 39-40). However, the ICLEI project coordinator admits that there are certain limitations with regard to learning from best practise case studies as these often describe a good idea without explaining how to transfer that good idea to another place (Interview with ICLEI, 08-08-2012).

As for Nagpur and Freiburg's respective contexts for clean energy development, ICLEI realised after the exchange visits that the cities' conditions were too different to directly transfer policies and practises between them. An ICLEI employee points out that a key difference between German and Indian cities is that while German cities often have authority over energy distribution, in India this is a state matter, so cities can only focus on demand-side management and not on energy supply. As a result, ICLEI focused on concrete project implementation in the areas of awareness-raising initiatives as well as energy efficiency and small-scale renewable energy demonstration facilities which were tailored to the responsibilities of Indian cities, rather than drawn directly from the European Resource cities (Interview with ICLEI, 05-12-2013).

Systematic Cooperation, Continuity of Interaction and External Moderator/Facilitator

As a consequence of the shift towards prioritising project implementation in the Indian Model Communities, ICLEI did not facilitate a systematic approach of direct knowledge exchange and learning between Nagpur and Freiburg. The direct exchange remained limited to the two mutual exchange visits during which officials presented their respective energy strategy and achievements to the other party. The Freiburg representative highlights that the ICLEI workshop as part of his visit to Nagpur could have been a promising starting point for more comprehensive exchange and he considers the fact that ICLEI did not follow up on this initiative as a missed opportunity.

He points out that the Nagpur representatives also showed interest in intensifying the exchange and he regrets that the interaction did not continue (Interview with Freiburg city administration, 07-08-2012).

Continuous direct exchange between Freiburg and Nagpur thus did not take place during any phase of the LRMCN program. Despite the focus on project work in the Indian cities, ICLEI actually tried to facilitate more personal exchange and regularly invited all LRMCN member cities to international conferences such as the UNFCCC Conferences of Parties and the annual Local Renewables Conferences in Freiburg. However due to visa problems and time constraints representatives from the two cities never managed to join any of these meetings at the same time.

Drawing on its learnings from the exchange with Nagpur and other international partnerships that did not lead to concrete project cooperation, Freiburg now plans to put more structure into its future exchange activities. Referring to a recent request of Tel Aviv to become Freiburg's partner, the member of the Freiburg city administration states that Freiburg is only willing to accept this invitation if there is the potential for intensive, systematic personal exchange and actual scope for joint project work (ibid.).

While generally Freiburg has established a wide network of international cooperation which the city largely manages without external support and moderation, the partnership with Nagpur was primarily driven by ICLEI. Similar to the GIZ's role in the Nashik-Hamburg collaboration, ICLEI initiated, facilitated and moderated all exchanges between the partner cities as part of the LRMCN. ICLEI set up the Local Renewables Initiative and the LRMCN network and selected all member cities based on their expressions of interest, socio-economic profiles and prior experiences with clean energy and environmental projects. ICLEI also organised all network events such as exchange visits, peer-to-peer meetings and workshops and moderated these events to bridge the different cultures and languages (see section on 'Intercultural Competence and Communication' below for more details).

ICLEI also played a key role in the implementation process of the LRMCN activities in Nagpur. ICLEI's regional office in Delhi staffed the Resource Centre in Nagpur with one permanent employee. This employee gave constant input on the setting up of the clean energy strategy in Nagpur throughout the entire LRMCN project period. She was in charge of conducting and updating the energy and GHG emissions inventory and she was also instrumental in developing and implementing project proposals, drafting the Renewable Energy and Energy Efficiency Policy and organising the stakeholder dialogue.

A Nagpur MC employee involved in the project confirms that the cooperation with ICLEI as part of the LRMCN has been rewarding for Nagpur. He highlights two areas where the city has particularly benefited from ICLEI's support, namely the energy inventory and international experiences. He is very thankful that ICLEI provided him with the opportunity to visit other cities in India and abroad to learn from their experiences and to present Nagpur's achievements at domestic and international conferences and workshops. He emphasises that without ICLEI Nagpur would not have received such international exposure, something which was instrumental for developing a vision for Nagpur's renewable energy strategy (Interview with Nagpur MC, 14-10-2013).

At the same time, ICLEI's leadership in the LRMCN also led to a dependency on the city network both in the partnership exchange and in the project implementation in Nagpur. This is exemplified in the energy and GHG emissions inventory which was set up and annually updated by ICLEI. The final project report states: "ICLEI's role during this step of the project was crucial, as it enabled the city governments to undertake these large exercises without building extra technical capacity within their governments." (ICLEI 2010b, 31) However, the downside of the inventory being created and managed by an external body was that when the project was completed Nagpur's city administration lacked the capacity to keep it updated.

Furthermore, the fact that ICLEI was also highly influential in preparing and implementing the activities led to challenges building local ownership of the LRMCN projects in Nagpur. ICLEI itself was aware of this challenge: "ICLEI South Asia was the main facilitator of the project activities. However, it was important that the cities accept ownership of the project through official endorsement and approval of the activities." (ibid., 33) While the collaboration between ICLEI staff and the Nagpur MC staff worked well in the LRMCN projects, the problem of lacking ownership has become more apparent in the more recent Solar Cities Program. The responsibility for the Solar Cities Program has been transferred to a different Nagpur MC employee who is generally more sceptical towards the collaboration with ICLEI. In an interview he complains that ICLEI employees tell him what to do without having adequate expertise (Interview with Nagpur MC, 19-11-2012). ICLEI employees also acknowledge that the cooperation with the Nagpur MC ran more smoothly in the LRMCN than in the Solar Cities Program (Interview with ICLEI, 05-12-2013).

Policy Windows and Intrinsic Interests

Similar to the other three case studies, the partnership between Nagpur and Freiburg as part of the LRMCN did not benefit from a clear policy window of opportunity. The political conditions were neither generally favourable nor overall adverse. The demand for innovative strategies to improve urban energy systems (the problem stream) is clearly visible in the rapidly growing city of Nagpur. Like many other Indian cities, Nagpur suffers from energy shortages and an overburdened electricity distribution infrastructure. The lack of energy security was identified as a key problem in Nagpur during the stakeholder consultations as part of the LRMCN, and Nagpur's Commissioner Chandra highlighted renewable energy sources and energy conservation as crucial alternative methods to address this problem (Chandra, 2006).

The availability of adequate solutions to develop clean energies at the urban level (the policy stream) has been rather limited. Nagpur was among the first cities in India to engage in city-level renewable energy and energy efficiency projects and could not draw widely from other Indian cities' experiences in this area. The LRMCN final project report confirms that Indian cities are still searching for innovative energy approaches: "The present time is right to develop more such programs, as the cities are beginning to invest their time and money to meet the energy needs of the present as well as planning for future growth. The cities are looking for support and advice, as they don't want to go in the wrong direction or 'reinvent the wheel'." (ICLEI 2010b, 54-55) Over the course of the project's implementation in Nagpur ICLEI referred to the experiences from the European Resource Cities as well as ICLEI's other member cities. But a systematic adoption of existing policy solutions from other cities did not take place as ICLEI decided to prioritise local project work rather than international exchange.

As for political support for the partnership project (the political stream) the interest of the Indian government in urban clean energy development developed only during the course of the LRMCN project. An ICLEI employee points out that the topic of urban renewable energies was first added to the Ministry of New and Renewable Energies' agenda while ICLEI was implementing the LRMCN in the Indian cities (Interview with ICLEI, 05-12-2013).

She reports that at the local level political support for the LRMCN program was stronger at the beginning, when Commissioner Chandra in particular became engaged in the initiative, while some of his successors were less committed to the project (ibid., see also section on Leadership of State Actors below).

ICLEI deliberately tailored the LRMCN around the intrinsic interests of the Indian Model Communities such as Nagpur, highlighting the project's co-benefits of cost-saving and improved energy security for the municipal corporations (ICLEI 2010a, 51). An ICLEI employee highlights that the promotion of co-benefits is a precondition for any low-carbon initiative in an Indian city. She explains that since ICLEI opened its regional office in India in 2001 awareness of the causes and consequences of climate change has slowly increased in Indian cities, in part due to the introduction of the National Action Plan on Climate Change (NAPCC) by the national government.

But even for progressive cities such as Nagpur, the main focus still remains on improving urban services while climate mitigation is only of secondary concern (Interview with ICLEI, 31-10-2012). Indian cities thus regard climate mitigation as a co-benefit that accompanies a more efficient provision of urban services.

Intrinsic interests were however only addressed in Nagpur as the Freiburg representative was not able to relate Freiburg's ongoing initiatives to the LRMCN. The LRMCN design did not generally fit with Freiburg's priority of orienting its international cooperation activities towards more mutuality and concrete project work. The member of Freiburg's city administration explains that Freiburg wants to develop partnerships that go beyond the mere exchange of presentations about best practise projects. He highlights the need for an in-depth analysis of both partners' contexts and way of working as a crucial prerequisite for meaningful exchange. According to his experience, the need for more concrete and insightful exchange beyond mere best practise presentations is increasingly acknowledged by many cities (Interview with Freiburg city administration, 07-08-2012)

Nagpur's priorities towards international cooperation partly complement and partly oppose Freiburg's aim of developing projects with mutual benefits. A Nagpur MC employee points out that he has a strong personal interest in any kind of international contact and exchange as this motivates and inspires him in his work in Nagpur and often also helps to convince local decision makers of innovative ideas. At the same time he explains that many people in Nagpur are primarily interested in international partnerships that bring in funding rather than being interested in practical knowledge exchange (Interview with Nagpur MC, 14-10-2013).

5.4.3.2 Linkages between the Partnership Project and State Institutions

The partnership between Nagpur and Freiburg, similar to that between Nashik and Hamburg, exemplifies an externally-led, top-down approach to city cooperation. As for the second explanatory variable, the linkages with state institutions, the exchange between Nagpur and Freiburg receives the second highest total score of the four analysed cases in this study's index (only the Nashik-Hamburg cooperation scores higher). The links to state institutions are particularly strong at the local level in Nagpur as well as with regard to the project being funded via public resources. In Freiburg however the project is not formally embedded into local state institutions (see table 19).

Indicators: State Institutionalisation	Evaluation: LRMCN		
	Nagpur	Freiburg	Total
Formal Approval by Local State Institutions	++	--	+-
Formal Approval by (Sub)National State Institutions	+-	+-	+-
Public Human Resources	+-	+-	+-
Public Financial Resources	++	++	++
Public Coordinator	+-	+-	+-
Commitment of Local State Leaders	+-	+-	+-

Table 19: Partnership Project LRMCN: Findings from Index State Institutionalisation

Formal Approval by Local and (Sub)National State Institutions

One major target of the LRMCN was the strengthening of the role of local governments in developing clean energy in the Indian and Brazilian model communities. The LRMCN final report highlights that: "The municipal corporations of the model communities and the Satellite cities were the main actors from the city to be involved in project implementation." (ICLEI 2010b, 19). Consequently, the project implementation Nagpur was closely tailored to the local government's decision-making competences in the areas of renewable energy and energy efficiency. Because in India electricity production and distribution fall under state government authority, ICLEI and the Nagpur MC decided to focus the LRMCN activities primarily on small demonstration projects, raising awareness and demand-side management which are under the legal jurisdiction of the local government. An ICLEI employee explains that ICLEI opted for an early realisation of demonstration projects in order to showcase the short-term benefits of clean energy projects. According to this interviewee the pilot projects were crucial, as local decision makers regularly change and new political leaders needed to be convinced of the value of the LRMCN program (Interview with ICLEI, 05-12-2013). She explains that the project implementation was designed to closely involve the local governments of the Indian Model Communities, so as to avoid the reliance on Indian state and central level government approvals (Interview with ICLEI, 05-12-2013). In April 2006 Nagpur's local council, Nagpur MC and ICLEI signed a

5.4 Case Study Nagur – Freiburg: The Local Renewables Model Community Network   187

MoU to develop a city-level clean energy policy. The policy only covers "interventions in the sectors that are under their [the local government's] jurisdiction such as street lighting, water pumping, buildings, waste management, lighting in public areas, public transport, citizen's awareness, etc." (ICLEI 2007, 9) and was, according to the ICLEI employee, formulated as an open-ended commitment that did not require state government approval (Interview with ICLEI, 05-12-2013). The policy was formally adopted by the city council in August 2007 which ICLEI again considered as a means to overcome political leadership change: "This was an important part of the process as council ratification ensures stability of the policy irrespective of the political situation of the local government." (ICLEI 2010a, 25)

Despite the focus on local governments and the deliberate bypassing of higher-level state approval in most of the LRMCN projects, the two larger-scale projects were implemented in collaboration with Indian subnational and national state institutions. The water and energy efficiency audit in local water treatment plants and pumping stations was conducted in collaboration with the Indian government's JNNURM program and the wastewater reuse system for the local thermal power plant was built together with the JNNURM and the state electricity supplier, Mahagenco. Also the demonstration and awareness projects that ICLEI facilitated in Nagpur were embedded in higher-level state institutions, but only on the German side. All LRMCN activities were approved in advance by the project funder BMZ/GTZ and ICLEI submitted quarterly activity reports about the projects' progress to the GTZ (ICLEI in email correspondence, 26-09-2014).

In the city of Freiburg the LRMCN was not formally institutionalised at all. The city did not realise any projects under the program, no MoUs were signed, no policy was introduced and no legislation was changed as a result of the participation in the LRMCN (Interview with Freiburg city administration, 07-08-2012).

Public Human and Financial Resources

As for the involvement of state resources the partnership scores highly in both Nagpur and Freiburg.

The partnership is executed by state employees from both cities. According to the LRMCN final report on Nagpur, two Nagpur MC employees, the executive engineer and the deputy engineer, worked on the implementation of the LRMCN projects as members of the RC (ICLEI 2010a, 36). The head of the Energy Unit at the Environmental Protection Department in Freiburg's city administration was the only actor in Freiburg directly involved in the LRMCN program.

The LRMCN program was not purely led by state actors. The project management and coordination was conducted by ICLEI. This city network cannot be clearly defined as a state or private organisation, but rather as a new form of governance that is simultaneously state and non-state, global and local (Betsill and Bulkeley 2006). The final report lists up to 12 ICLEI employees that have worked on the LRMCN initiative: two to three employees at the ICLEI International Training Centre, conducting the project management, one to two employees from the European regional office that were involved in exchange and knowledge transfer, four employees from ICLEI South Asia acting as executing partners, plus one ICLEI staff member in each of the three Indian model communities' Resource Centres (ICLEI 2010b, 19).

Other than the financing of personnel, all of the LRMCN program's financial resources were provided by state institutions. Most LRMCN activities, such as the pilot projects, the energy inventories, the stakeholder meetings and some of the traveling by the Nagpur MC staff were realised with funding from the BMZ that was administered by the GTZ.

Nagpur's city administration also contributed resources, such as space for the Resource Centre, staff time and some funds for projects (Interview with ICLEI, 05-12-2013). The LRMCN final report for Nagpur specifies that Nagpur MC made financial contributions to short-term and awareness-raising projects, e.g. the solar lighting project at the Nagpur MC premises was fully funded by the Nagpur MC. Furthermore, the funding for the larger scale water sector audit and water reuse projects was provided by the JNNURM and the Maharashtra state electricity company Mahagenco (ICLEI 2010a; ICLEI 2010c).

Public Coordinator and Commitment of Local State Leaders

As elaborated above in the section on "Knowledge Transfer Process" ICLEI fulfilled a similar leadership role as external facilitator in the partnership between Nagpur and Freiburg as the GIZ played in the waste to energy project involving Nashik and Hamburg. ICLEI initiated the LRMCN and moderated all exchange between Nagpur and Freiburg and it coordinated the program implementation in Nagpur from the local RC and its regional office in Delhi. Other than the GIZ, ICLEI can however not be clearly defined as a state organisation as it also embodies characteristics and functions of a non-governmental organisation.

While ICLEI was a key driver of the LRMCN, the leadership and commitment of local state actors was more ambiguous. In the beginning of the project the Commissioner Chandra and the Nagpur city administration strongly supported the LRMCN program and the local RC (Interview with ICLEI, 31-10-2012). A Nagpur MC employee confirms that local political leaders were generally supportive and pro-active at the start of the LRMCN partnership, which was a requirement for receiving approvals quickly (Interview with Nagpur MC, 14-10-2013).

However, over the course of the project changing leadership has still been a key challenge in Nagpur, as several of Commissioner Chandra's successors were less convinced of the LRMCN's feasibility and benefits for Nagpur, which made it difficult for ICLEI to implement additional projects (Interview with ICLEI, 05-12-2013).

According to an ICLEI employee the involvement of the engaged city engineers from the Nagpur MC was crucial to continue the project in times of leadership changes, as these employees served as the organisational memory of the city administration and kept on proactively communicating and collaborating with ICLEI despite a leadership change (ibid.).

Freiburg's political leaders are generally highly committed to clean energy development. The city's Mayor, Dieter Salomon, is member of the Green Party and well known in and beyond Freiburg for developing and promoting it as an environmental model city. An ICLEI employee highlights that Salomon is particularly committed to pursuing and reviewing local clean energy development in Freiburg (Interview with ICLEI, 08-08-2012). A member of Freiburg's section of the environmental NGO Bund für Umwelt und Naturschutz (BUND) confirms that there is general consensus between Salomon and the city council on the issue of climate protection and clean energy projects in Freiburg (Interview with NGO BUND, 07-08-2012). An ICLEI employee emphasises that Freiburg also excels as one of ICLEI's most engaged member cities. She explains that Freiburg has shown on-going commitment to hosting the Local Renewables Conference and has always been willing to welcome international delegations such as the Nagpur representatives (Interview with ICLEI, 05-12-2013). Salomon also met with the Indian MNRE delegation on their visit to Freiburg in June 2007 (ICLEI 2007). Aside from this, however, political leaders from Freiburg have not played a major role in the LRMCN partnership.

5.4.3.3 Local Partnership Entrepreneurs

The involvement of engaged local individuals was limited in the exchange between Nagpur and Freiburg. Only the member of Freiburg city administration's Environmental Protection Department can be defined as a partnership entrepreneur. He was the main protagonist from Freiburg involved in the LRMCN, was personally engaged in pushing for the extension of the exchange with Nagpur, is well connected within and beyond the city of Freiburg and is an experienced coach in urban knowledge transfer (see table 20).

On the Nagpur side no partnership entrepreneur can be identified. The Nagpur MC city engineer who was ICLEI's main local partner in the LRMCN projects is the only individual who fulfils several criteria that could have theoretically enabled him to function as a partnership entrepreneur in the exchange with Freiburg. He has been an engaged and well-networked advocate for clean energy development within Nagpur's city administration. An ICLEI employee states that the city engineer and his team took ownership of the initiative, despite being burdened with many other responsibilities and duties (Interview with ICLEI, 05-12-2013). She further highlights that the LRMCN program benefited strongly from the city engineer's ability to connect several energy-related sectors of urban development. This was due to his responsibility over Nagpur's JNNURM projects which also combine several sectors such as water, waste management and energy efficiency (ibid.).

The Nagpur city engineer was mainly ICLEI's partner and he was not actively involved in the direct exchange with Freiburg. He can therefore not be classified as partnership entrepreneurs. As a result, Nagpur scores zero in this index (see table 20).

Indicators: Partnership Entrepreneur	Evaluation: LRMCN		
	Nagpur	Freiburg	Total
Belief in Feasibility and Benefits	--	++	+-
Investment of own Resources	--	--	--
Ability to Convince Stakeholders	--	+-	+--
Internal Policy Network	--	++	+-
External Policy Network	--	++	+-
Partner City Policy Network	--	--	--

Table 20: Partnership Project LRMCN: Findings from Index Partnership Entrepreneur

Belief in Feasibility and Benefits, Investment of Resources and Ability to Convince Stakeholders

At the beginning of the LRMCN program, the member of the Freiburg city administration showed great interest in fostering the exchange with Nagpur. As the head of the Energy Unit in the Environmental Protection Department he has promoted local green energy development for more than 20 years. He is strongly convinced of the benefits of renewable energies and energy efficiency and stresses the importance of more horizontal learning between cities, even if he also highlights the fact that such knowledge transfer processes are complex and require extensive first-hand experience (Interview with Freiburg city administration, 07-08-2012).

In recent years he has established his own private consultancy and offers cities in Germany and abroad support preparing and implementing local energy and climate programs. He considered the exchange with Nagpur as a promising start to a fruitful future cooperation and he was astonished that the exchange with his partners from Nagpur stopped suddenly after the initial visits. He made several attempts to re-establish contact by email, but his contacts in Nagpur did not respond and he eventually gave up.

As for the input of personal resources, the Freiburg city administration employee regularly invests his free time in urban exchange projects. For example he works as a coach supporting smaller municipalities in the setting up of a local climate and energy strategy as part of a pilot project run by the city network Climate Alliance called "Coaching Kommunaler Klimaschutz".[54] As part of the Nagpur exchange he was however officially released from his role at the Freiburg city administration for one week in order to visit Nagpur, the costs of which were reimbursed by the GTZ.

Due to his work in Freiburg and other German and international cities he has wide experience collaborating with different public and private stakeholders that play important roles in the realisation of climate and energy strategies. He was selected as one of the Climate Alliance's coaches and ICLEI staff also regularly invite him to speak on city-level topics at ICLEI events being a charismatic advocate for local clean energy development (Interview with ICLEI, 08-08-2012). In the LRMCN his direct interaction with local stakeholders from Nagpur was limited to one exchange visit and he did not have the opportunity to interact with them again directly at later stages of the project.

Internal, External and Partner City Network

At the time of the interview for this study (August 2012), the Freiburg representative had lost contact with his peers and project partners from Nagpur.

Generally, he has very good access to relevant policy networks in the areas of climate protection and energy management, both in Freiburg and beyond. In Freiburg he is well connected to local decision makers such as Mayor Salomon through his function as head of the Energy Unit in the Environmental Protection

---

54   http://coaching-kommunaler-klimaschutz.net/ (19-02-2016)

Department of Freiburg's city administration. He is instrumental in setting Freiburg's climate policy as well as managing Freiburg's international contacts in the area of climate and energy and he communicates directly with Salomon and the Freiburg city council on these issues. Furthermore, he serves as a member of the Sustainability Council in Freiburg. The council is a non-partisan board made up of Salomon, local decision makers from all major parties, private experts and stakeholders as well as other members of the city administration that advise Freiburg on the implementation of the city's sustainability targets.[55]

He also has access to an extensive network of contacts outside of Freiburg. He points out that he has regularly contact with his colleagues from other environmentally-progressive German cities such as Heidelberg and Munich when he is confronted with problems and challenges that Freiburg has not dealt with before, knowing that other cities have (Interview with Freiburg city administration, 07-08-2012)

He adds that the majority of his contacts is with cities of a similar size to Freiburg and concludes that he seldom approaches smaller municipalities as their contexts, resources and issue areas differ too greatly from Freiburg's (ibid.)

The Freiburg representative also regularly interacts with partners from cities outside of Germany, whether with Freiburg's sister cities such as Besancon (France), Padua (Italy), or Madison (USA) or via Freiburg's involvement in the city networks ICLEI, the Climate Alliance and Energy Cities. He also participated in a GIZ project on energy management in capital cities in the Balkans, such as Zagreb, Podgorica, Sarajevo, Skopje, Belgrade and Tirana (ibid.).

5.4.3.4 Partnership Social Capital

As the direct interaction between the actors from Freiburg and Nagpur in the LRMCN program was confined to the initial exchange visits, the opportunities to develop social capital were very limited. The Nagpur-Freiburg exchange actually scores lowest of the four cases analysed in the partnership social capital index. It does not gain a high score in any of the eight indicators and receives particularly low scores for prior partnership experience, trust, equality and intercultural communication (see table 21).

---

55 http://www.freiburg.de/pb/,Lde/206104.html (29-04-2014)

5.4 Case Study Nagur – Freiburg: The Local Renewables Model Community Network

Indicators: Partnership Social Capital	Evaluation: LRMCN		
	Nagpur	Freiburg	Total
Prior Collective Action	/	/	--
Trust	--	--	--
(Perceived) Equality	--	--	--
Intercultural Experiences	+-	+-	+-
Intercultural Communication	/	/	--
Involvement of Leadership Networks	++	--	+-
Inclusion & Citizen Participation	+-	--	+--

Table 21: Partnership Project LRMCN: Findings from Index Partnership Social Capital

Prior Collective Action, Trust and (Perceived) Equality

Nagpur and Freiburg had no direct contact before ICLEI brought them together for the LRMCN program and could thus not draw on any experiences and relations from prior partnership work. Both cities were already members of ICLEI when the LRMCN started in 2005 but the city network operates via regional offices (Freiburg is part of ICLEI Europe and Nagpur is member of ICLEI South Asia) and works primarily within these regions. The LRMCN was one of ICLEI's first attempts to bring cities from the Global North and South together in one project (Interview with ICLEI, 05-12-2013).

As the direct exchange between Nagpur and Freiburg was also limited in the LRMCN, the protagonists involved did not develop personal, trusting relations which are a crucial component of any form of social capital. Similar to the reliance of Nashik and Hamburg on the GIZ, Nagpur and Freiburg were fully dependent on ICLEI to facilitate and moderate the exchange. In contrast to the GIZ, ICLEI in principal permitted direct communication between Nagpur and Freiburg outside of the official project framework and the Freiburg representative actually attempted to contact his counterparts in Nagpur several times to keep the exchange alive but he did not receive any response. The ICLEI project coordinator states that one of the reasons for this may have been that he lacked the personal connections to his peers in Nagpur which are of great importance for any kind of collaboration in India, whereas just sending emails is not sufficient to reach people (Interview with ICLEI, 08-08-2012)

As for the challenge of realising equality within urban North-South cooperation, the Freiburg representative expressed his interest in developing a partnership with Nagpur based on an equal footing, with both cities learning from each other (Interview with Freiburg city administration, 07-08-2012). However, from its conception onwards, the LRMCN contained certain elements that constrained the establishment of equality in the partner city relationship. The LRMNC to a certain degree reinforces the traditional one-directional learning process, with European cities guiding Indian and Brazilian cities in the establishment of energy strategies. An ICLEI employee involved explains that ICLEI placed the LRMCN member cities into three different categories; the European "Resource Cities" and the Indian and Brazilian "Model Communities" and "Satellite Cities", thereby entrenching the different roles of the cities in the project. The European "Resource Communities" were given the task to provide advice and guidance based on their existing experiences with clean energy development. The "Model Communities" are the cities in India and Brazil where ICLEI actually developed and implemented projects. The "Satellite Cities" were cities that observed the processes undertaken in the "Model Communities" in order to get inspired and learn from them (Interview with ICLEI, 31-10-2012). The Nagpur project report also emphasises the different roles of the LRMCN member cities, highlighting that the European "Resource Cities" acted as reference models and consultants for the Indian and Brazilian cities: "They offered advice, motivated change, provided relevant examples of policy and technology approaches, and shared reference projects with the Indian and Brazilian model communities" (ICLEI 2010a, 4).

A partnership based on an equal footing with mutual learning benefits was thus not a key aim of the project.

Intercultural Competence and Communication

As for the partnership social capital indicator of intercultural experiences, both Freiburg and Nagpur have conducted international partnership projects prior to the LRMCN. Freiburg has close relations with its sister cities, in particular Besancon and Padua, and the city has been a member of ICLEI and other transnational city networks since the early 1990s (Interview with Freiburg city administration, 07-08-2012). Nagpur, and in particular the city engineer that was ICLEI's main contact in the Nagpur MC, had also collaborated with international experts before. He refers to a water treatment project in 1997 where experts from Germany, France and Japan served as consultants and regularly visited Nagpur. He also visited a GIZ consultant in Bonn in 2004 (Interview with Nagpur MC, 14-10-2013).

Both the Nagpur MC city engineer and the member of Freiburg's city administration have primarily worked with partner cities from their own countries and for the Freiburg representative the exchange with Nagpur was his first contact with India. Even for ICLEI it was one of the first projects it had conducted that involved both German and Indian cities (Interview with ICLEI, 05-12-2013). The overall intercultural experience in German-Indian collaboration of the main partnership protagonists was thus limited.

The lack of in-depth knowledge of the partner cities' different cultures and habits led to major communication barriers in the partnership. The interaction between Nagpur and Freiburg worked relatively well as long as ICLEI organised and moderated the exchange. The ICLEI project coordinator states that one of ICLEI's key tasks in the project was to serve as a linguistic and cultural bridge between the cities. She explains that in particular the linguistic translations were very important in the face-to-face meetings as not everybody understood everybody else's English; but that there was also a need for cultural translations as sometimes the partners from Germany and India interpreted each other's statements wrongly (Interview with ICLEI, 08-08-2012).

The project coordinator adds that the use of email was one of the major differences in communication cultures between the Indian and European cities which made direct exchange difficult in the LRMCN. She explains that in Indian cities email communication among administrative peers is still uncommon as it does not fit well with the hierarchical structure of the city administrations where the most senior person has to be contacted first who then delegates the request to the person who deals with it. She adds that the Freiburg representative might have been more successful if he had involved ICLEI staff in his attempts to contact his colleagues from Nagpur as ICLEI could have then followed up the communication. She further states that the Freiburg representative may have wrongly interpreted the fact that he did not get a reply as an indicator for lacking interest to continue the exchange on the side of Nagpur (ibid.)

Another ICLEI employee involved confirms that despite improvements in Indian cities with regard to email communication in recent years the different email cultures have been a major barrier for more direct exchange between Nagpur and Freiburg. She states that the lack of using emails in Indian cities makes it difficult to set up a joint virtual platform with foreign partner cities (Interview with ICLEI, 31-10-2012)

Involvement of Leadership Networks, Inclusion, and Citizen Participation

Local stakeholder networks and citizens from Freiburg did not play a major role in the LRMCN. The only stakeholder mentioned in the documents and interviews is Dr. Eicke Weber, the head of the Fraunhofer Institute for Solar Energy. He met with the MNRE delegation that visited Freiburg in June 2007 to discuss the potential for collaboration (ICLEI 2007). Weber was however not involved in the LRMCN beyond this visit.

In Nagpur, the involvement of local leadership groups and, to a lesser degree, inclusion was realised in the LRMCN program. ICLEI defined local stakeholder participation as one of the LRMCN's four major targets (see above) and designed the implementation process to involve stakeholders both as consultants and project partners. At the beginning of the project the Resource Centre invited local stakeholder groups to join a roundtable meeting in Nagpur to prepare the local energy policy which was adopted by the Nagpur city council a few months later (ICLEI, 2010a). According to an ICLEI employee the involvement of stakeholder groups was particularly important at this stage of action planning in order to widen the LRMCN's scope beyond the influence of the city government (Interview with ICLEI, 05-12-2013)

Local stakeholders also provided consultancy at later stages of the project for the reviewing of the annual energy reports and the Resource Centre's proposals for project activities (ICLEI 2010a). An ICLEI employee concludes that the involvement of stakeholder groups was crucial for three main reasons, namely: identifying the different responsibilities of relevant actors for local clean energy development; reaching out to non-governmental sectors in Nagpur; and continuously motivating the city government to progress with the project (Interview with ICLEI, 05-12-2013).

In addition to serving as consultants, local stakeholders were also involved as partners in the LRMCN awareness and demonstration projects, both as co-organisers and participants. For example local women's groups, school classes and local citizens were trained at several workshops in energy saving and renewable energy use. The school training was prepared in collaboration with the local environmental NGO "Vidharbha Akshay Urja Vikas Probodhini" (ICLEI 2010a). The RC also contracted local renewable energy businesses to execute pilot projects and encouraged them to display their products in its Resource Centre facilities (ICLEI 2010b).

While the inclusion of marginalised groups within the local population was not a defined priority in the LRMCN, the initiative did involve local citizens more broadly in the projects. The awareness-raising and demonstration projects in particular aimed to reach members of the local population and introduce as many

Nagpur citizens as possible to the novel topic of local renewable energy and energy efficiency development. In two of the LRMCN projects the poorer population is explicitly mentioned as the target group, the first being the solar water heater facility which was built in the municipal hospital: "The installation will act as a pilot project to demonstrate advantages of using solar technologies and provided added health benefits to the visitors of the hospitals, mostly poor urban dwellers." (ICLEI 2008, 4) Also the training programs on renewable energy and energy efficiency for school children were primarily conducted in municipal schools that mainly educated pupils from less wealthy families in India (ICLEI 2010a). Furthermore, between December 29 and 30, 2007 the RC in collaboration with the Indian Institute of Youth Welfare held a workshop on energy efficiency for local women's groups which was attended by 42 participants (ibid.).

6 Cross-Case Comparison: Testing the Research Hypotheses

6.1 Results from Comparative Case Study Analysis

As elaborated in chapter 4.2 the comparative perspective provided by multiple case studies allows for the identification of patterns in the independent variables of the phenomenon analysed. This is a crucial prerequisite for the analytical generalisation of the findings in qualitative research designs (Yin, 2009). In this comparative case study analysis four independent variables have been drawn from the literature to derive explanations for success and failure in transnational urban partnership projects on low carbon development. The independent variables have been translated into the following four hypotheses:

Hypothesis 1: A transnational urban partnership project is more likely to succeed if it follows a well-prepared knowledge exchange strategy.

Hypothesis 2: The more a transnational urban partnership project is institutionalised into the state system, the more likely the project is to succeed.

Hypothesis 3: A transnational urban partnership project is more likely to succeed if it is driven by engaged, persuasive and well-networked partnership entrepreneurs.

Hypothesis 4: The more social capital protagonists develop as part of the transnational urban partnership, the more likely the partnership project is to succeed.

An index system has been developed to operationalise the hypotheses. The index system includes one index for each independent variable plus a success index measuring the dependent variable. The four partnership projects have been scored according to this Index System, based on the evidence of expert interviews and document analysis. Table 22 shows the results of the index scoring.

	Success Index (total 14)	I1: Exchange Strategy (total 14)	I2: State Institutionalisation (total 12)	I3: Partnership Entrepreneur (total 12)	I4: Social Capital (total 14)
DEWATS	6	7.5	2.5	10	8.5
Tramway	4	7.5	3.5	10	6.5
W2E (tentative)	7	11	9	0	3
LRMCN	7.5	6	7	3.5	2.5

Table 22: Results of the Index System Scoring (Success Index highlighted)

6.2 Comparing Partnership Outcome: Mixed Success and Differing Strengths and Weaknesses

Starting with the dependent variable, the success index reveals that the four partnership projects scored between 4 (low success) and 7.5 (medium success) out of a total 14 points that can be gained in this index measuring project success. The LRMCN project (score 7.5) and the W2E project (tentative score 7) gain just or a little more than just half of the 14 points possible. The other two partnership initiatives score lower in the success index, the DEWATS project gaining a score of 6 and the tramway transfer initiative a score of 4. It is again worth noting that the assessment of the W2E project is tentative as the project had not been completed by the end of the data collection period for this study (see chapter 5.3). What can however be said is that at the time of this study's data collection none of the projects analysed was assessed to be fully or even largely successful according to the indicators set up in the success index. But none of the projects was shown to be a complete failure, either.

The comparison of scorings of individual success indicators provides more detailed insights into the projects' differing outcomes (see table 23). As described in chapter 4.4, a high indicator score (++) receives two points, a medium score (+-) one point and a low score (--) no points. Half points account for differing indicator scores between the partner cities, e.g. the indicator "Capacity Building" in the LRMCN project scores a total of 0.5 (+--) as the score for Nagpur is 1 and for Freiburg 0. The accumulated scoring of the seven indicators results in the projects' index scoring.

6.2 Comparing Partnership Outcome: Mixed Success and Differing Strengths and Weaknesses 201

	Project Implementation	Budget & Schedule	Capacity Building	Benefits for Target Groups	Project Impact	Mutuality	Long-term Partnership	Success Index (total 14)
DEWATS	+-	+-	+-	+-	+-	--	+-	6
Tramway	--	--	+-	--	--	++	+-	4
W2E (tentative)	++	--	+-	+-	+-	++	--	7
LRMCN	++	+-	+--	++	++	--	--	7.5

Table 23: Success Index: Results of the Indicator Scoring

As for the first success indicator that assesses a project's actual implementation, the LRMCN partnership scores highest, as the project partners realised several pilot, awareness and large-scale projects in Nagpur as part of this initiative. Also the W2E project receives tentatively high scores, as the successful implementation of the project seemed very likely at the time of this study's completion. In the DEWATS partnership only the NGO-led pilot projects were successfully implemented, while the wider dissemination of DEWATS in Pune failed. The tramway initiative has not been implemented and it is unlikely that the tramway transfer will ever be realised.

All four partnership projects faced difficulties with regard to their performance in terms of adherence to their budget and/or schedule, the second success indicator. None of the four initiatives fully kept to their initial time schedule but the extent of the delays they incurred differs. The DEWATS and LRMCN initiatives saw delays to the implementation of some of their projects of up to a few months (the BORDA-MAM plant in Hadapsar, part of the DEWATS partnership and several renewable energy and energy efficiency projects that were part of the LRMCN project). The PMC-run DEWATS plant and the W2E project have been delayed for several years and the tramway initiative is unlikely ever to be completed. The tramway partnership project also faced the most serious problems with its budget, as the project funders, the city administrations of Pune and Pimpri Chinchwad, refused to pay a significant part of the stipulated sum for the completed DPR to the German-Indian team of consultants. In the W2E project, the time delays forced the project coordinator, the GIZ, to make cuts to the budget. The DEWATS plants (particularly those in Devrukh and Narodi) and the LRMCN project performed well in terms of their budget.

The third indicator in the success index, local capacity building, measures improvements in the areas of learning by individuals, administrative processes, institutional reform and establishment of local knowledge centres. This indicator has been scored separately for both the German and Indian cities in the partnership projects (see table 24).

Indicator: Capacity Building			
	Indian City Score	German City Score	Total Score
DEWATS	+-	+-	+-
Tramway	+-	+-	+-
W2E (tentative)	+-	+-	+-
LRMCN	+-	--	+--

Table 24: Indicator Capacity Building: Scoring of the Indian and German Cities

With the exception of Freiburg that saw no capacity building take place during the LRMCN project, the project actors from all the other cities involved in the four partnership projects realised capacity building, at the very least in the form of individual learning. Capacity building in the form of improved administrative processes and institutional reform was less evident, taking place only in Nagpur and to some degree in Pune and Pimpri Chinchwad as part of the tramway project. Nagpur introduced energy and GHG emissions inventories and a local energy policy as part of the LRMCN project. Despite its failed implementation, the tramway project also had a positive effect on the way transport was managed in Pune and the neighbouring city of Pimpri Chinchwad, as discussions about a joint tramway system led to the merger between the cities' transport departments. Moreover, the tramway alignment map has served as a basis for the planning of Pune's metro system. In Nashik the institutional framework and administrative processes required for the waste-to-energy project were already largely in place when the W2E project started and this was in fact the main reason why the city was selected as the project's site. No institutional and administrative reforms were undertaken in Pune during the DEWATS project or in any of the German partner cities. Hamburg Wasser however aims to facilitate more integrated waste and wastewater management in Hamburg and other German cities by showcasing the example of the waste-to-energy project in Nashik, once the plant is built.

6.2 Comparing Partnership Outcome: Mixed Success and Differing Strengths and Weaknesses   203

Local knowledge centres were set up in Nagpur (the Resource Centre) and Pune (the International Office Agenda 21). While the IOA 21 remained largely ineffective, the RC was the major institution leading and guiding the LRMCN project's implementation. However, the RC has remained strongly dependent on the external project coordinator, ICLEI, to realise projects and influence local decision makers.

Only Nagpur from the LRMCN project scores highly in the fourth success indicator measuring the concrete benefits for target groups provided by the projects (see table 23). Thanks to its participation in the LRMCN project Nagpur's city administration, the LRMCN's main target group, received funds for local energy projects, plus energy and emissions inventories as well as national and international exposure and recognition. In addition local solar energy businesses improved their sales and school children, the third target group, took part in projects which developed their knowledge of clean energy sources. In the DEWATS and W2E initiatives the target groups partially benefited from the projects. In Pune the NGO-led DEWATS pilot plants are still operating well and the facilities they support receive efficient wastewater treatment with low maintenance levels. Beyond these demonstration plants however, the majority of Pune's population has not seen any improvements due to the initiative, due to the failure of the PMC plant for former slum dwellers and the lacking wider diffusion of the DEWATS approach in Pune. In Nashik the citizens that use the public toilets already benefit from the partnership between Nashik and Hamburg Wasser, even if the waste-to-energy project has yet to be completed. Recommendations from Hamburg Wasser employees on how to improve local septic tank management have been implemented, improving hygiene conditions for the local population. As the tramway system was never built in Pune, the project's target groups, local citizens and in particular commuters, do not enjoy any benefits from the partnership initiative.

The fifth indicator, the wider impact of the partnership projects beyond the partnership initiatives, has received similar scores to the fourth indicator; the LRMCN initiative scores highest, the DEWATS and W2E partnerships receive a medium score and the tramway project scores lowest in this indicator. The LRMCN initiative has had a national impact in India, which in fact exceeded the expectations of the project coordinator ICLEI. The LRMCN and the city of Nagpur were selected to serve as models for the design of the Government of India's Solar Cities Program which has currently been implemented in 60 Indian cities. Moreover, several Indian cities adopted Nagpur's energy and water audit management approach. The DEWATS project had some impact, as BORDA established an international DEWATS conference series as a result of the experiences in Pune and several of the pilot plants have served as demonstration models for outsiders, such as the Maharashtra State Minister and international students. As the GIZ will only start promoting the W2E project once it is completed, its wider impact has so far been limited. But GIZ employees highlight that the project has already received interest from city, state and central government officials in India, even though the plant has yet to be built. Apart from the capacity building in Pune mentioned above, the tramway project has had no wider impact in and beyond Pune.

While the LRMCN project scores highest in several success indicators (measuring implementation achievements, benefits for target groups and wider project impact) it scores low for perceived mutuality in the partnership, the sixth success indicator (see table 23). Only the Indian partner Nagpur benefited from this partnership, receiving support for the setting up and implementination of a local clean energy strategy. In Freiburg no LRMCN activities were conducted. Similarly the DEWATS initiative focused only on realising projects in the Indian city of Pune, while in the German partner city Bremen no projects were planned and realised. The tramway and W2E initiatives score higher with regard to perceived partnership mutuality. In the tramway project the Indian partners, the city administrations of Pune and Pimpri Chinchwad, agreed to provide the funding for the DPR study (although in the end they only paid the first installment). The German consultants joined the project hoping to also be contracted to implement the tramway in Pune and Pimpri and to establish a wider network of business relationships in India. The W2E partnership also scores relatively highly for perceived partnership mutuality as both Nashik city administration and Hamburg Wasser have clear self-interests in the project. Nashik receives a low cost solution for improving their local sanitation and waste collection systems and Hamburg Wasser gets the opportunity to test and demonstrate its HWC technology in the context of a city in the Global South.

The seventh success indicator measures the long-term development of the city partnerships (see table 23). The results reveal that the W2E and LRMCN projects were conducted in the form of ad hoc partnerships. The German and Indian city actors were only brought together for these projects and both partnerships are unlikely to survive beyond these single initiatives. The DEWATS and tramway partnership initiatives score higher in the assessment of their post-project sustainability. Both projects have been part of a long-term city partnership between Pune and Bremen which continues today, though it has been less active in recent years.

Comparing the total scores for the four case studies in the success index a negative correlation between two of the indicators can be identified; the first and the seventh indicator appear to have contrasting scores. Those two partnerships that have realised (LRMCN) or will likely realise (W2E) macro-level project implementation, receive very low scores for the development of long-term relations between the partner cities. The DEWATS and tramway initiatives on the other

hand have failed to achieve macro-level implementation, but they score better with regard to the long-term development of the city partnership.
This finding leads to the following proposition:

Proposition 1: There appears to be a trade-off between the realisation of large-scale partnership projects and the long-term development of transnational urban partnerships.

6.3 Explanatory Factors in a Comparative Perspective

In the following section, the four hypotheses of this study are tested by comparing the index scores for the independent variables with the results of the success index. On the basis of these results the author will evaluate whether the hypotheses provide plausible explanations for the differing outcomes of the four partnership initiatives.

6.3.1 Knowledge Exchange Strategy

Hypothesis 1: A transnational urban partnership project is more likely to succeed if it follows a well-prepared knowledge exchange strategy. The first hypothesis argues that the setting up of a comprehensive and well-prepared approach for realising knowledge transfer is crucial for the success of urban partnership projects. Table 25 highlights the results of Index 1 that measures the quality of the projects' knowledge exchange strategies.

The index scores range from 6 (low to medium quality of the exchange strategy) in the LRMCN initiative to 11 (high quality of the exchange strategy) in the W2E initiative, out of 14 points possible in this index.

The two projects conducted as part of the Pune-Bremen partnership achieve similar, medium scores in the index measuring their knowledge exchange strategies. There is no apparent correlation with their scores in the success index which are relatively low in both cases, with the DEWATS project scoring slightly better than the tramway initiative.

	Success Index (total 14)	I1: Exchange Strategy (total 14)	I2: State Institutionalisation (total 12)	I3: Partnership Entrepreneur (total 12)	I4: Social Capital (total 14)
DEWATS	6	7.5	2.5	10	8.5
Tramway	4	7.5	3.5	10	6.5
W2E (tentative)	7	11	9	0	3
LRMCN	7.5	6	7	3.5	2.5

Table 25: Results of the Index System Scoring (Index 1 "Exchange Strategy" highlighted)

The scorings of the two other partnership projects deliver some more unexpected results. In fact, the project with the highest score in the success index, the LRMCN initiative, scores lowest in the index measuring the quality of its knowledge exchange strategy, thereby questioning hypothesis 1 of this study. At first glance, the hypothesis appears to be further challenged by the fact that the W2E project scores highest of all four projects in the index "Exchange Strategy" while the project only receives a medium score in the success index. Both results should however be treated with care and their significance should not be overestimated. As explained above, the scoring of the W2E initiative is tentative as the project's implementation was still in progress at the time of the completion of the data collection for this study. A later assessment of the initiative could result in a better scoring in the success index. The low score for the LRMCN initiative primarily results from the fact that the direct exchange between Nagpur and Freiburg was limited and confined to the early project phase. The project coordinator ICLEI then decided to focus more on concrete project implementation in Nagpur based on indirect lesson-drawing from ICLEI's experiences with urban clean energy development in Freiburg and its other member cities, rather than developing a strategy to facilitate more direct personal exchange among the LRMCN network cities.

Based on the results of the index comparison even an analytical generalisation with regard to hypothesis 1 would be thus questionable.

However, the comparison of the individual indicators that make up the index "Exchange Strategy" reveals more tangible insights (see table 26).

	Program Evaluation	Context-Evaluation	Systematic Cooperation	Continuity of Interaction	External Moderation	Policy Window	Intrinsic Interests	I1: Exchange Strategy (total 14)
DEWATS	++	+-	+-	++	--	+-	+--	7.5
Tramway	++	+-	+-	+-	--	+-	++-	7.5
W2E (tentative)	++	++	+-	+-	++	+-	++	11
LRMCN	+-	+-	--	--	++	+-	+-	6

Table 26: Index 1 Exchange Strategy: Results of the Indicator Scoring

While the LRMCN's scores for the individual indicators in the "Exchange Strategy" index are exceptional due to the limited direct exchange between the cities, the indicator scores for the DEWATS, the tramway and the W2E projects are similar in many aspects. The three projects have all focused on an in-depth prior evaluation of the transferred technology, they have found it challenging to continuously follow a systematic approach in their cooperation and none of the projects has benefited from a clear policy window. The three projects also do not differ substantially with regard to the prior context evaluation and the continuity of direct personal interaction. In the comparison, no clear patterns link the results of these indicators to the actual success of the project.

Such a pattern can however be identified for the remaining two indicators. The first pattern is found in the indicator measuring the involvement of external partnership moderators. The DEWATS and tramway projects receive low scores and the W2E project scores highly in this indicator. The LRMCN project also scores highly for the involvement of external partners. In fact, the LRMCN project only scores high in this indicator while receiving low or medium scores for the remaining indicators measuring the quality of the partnership's exchange strategy. Comparing these results with individual success indicators reveals initial explanations for the earlier identified negative correlation between macro-level implementation success and post-project sustainability in the projects analysed. The two partnership initiatives driven by external moderators have realised (LRMCN) or will likely realise (W2E) larger scale projects, but it is unlikely that these city partnerships will continue after the projects. In contrast, the two projects that did not involve external moderators have failed to realise larger scale projects (tramway) or the city-wide dissemination of the technology beyond pilot projects (DEWATS). But the Pune-Bremen city partnership, under which both the DEWATS and tramway initiatives have been set up, has continued beyond the projects. According to these results, the involvement of external moderators tends to support the implementation of large-scale partnership projects, but at the same time it appears to hinder the long-term development of partnerships.

A second relationship can be found between the indicator "Intrinsic Interests" and the success indicator "(Perceived) Mutuality". The indicator "Intrinsic Interests" has been scored separately in both the Indian and German partner cities (see table 27). The scorings reveal that the two initiatives that incorporated the intrinsic interests of both the Indian and German partner cities (tramway and W2E) also score highly for the success indicator "(Perceived) Mutuality", whereas the two initiatives that only focused on addressing the interests of the Indian partner city (DEWATS and LRMCN) receive low scores for perceived partnership mutuality.

Indicator Intrinsic Interests		
	Indian City Score	German City Score
DEWATS	+-	--
Tramway	+	++
W2E (tentative)	++	++
LRMCN	++	--

Table 27: Indicator Intrinsic Interests: Scoring of the Indian and German Cities

Summing up the results of the index "Exchange Strategy", the four cases' total scores in the index are too diverse to support the argument for a well-prepared and comprehensive knowledge exchange strategy in transnational urban partnership projects. This study's first hypothesis is thus neither confirmed nor rejected by the comparative analysis of the four case studies. However, the comparison of individual indicators has revealed several instructive patterns and proven to be a promising approach to refining the hypothesis. Firstly, external exchange moderators appear to play a key role in transnational urban partnerships, with the projects involving external moderators scoring better for the implementation of larger-scale projects and the projects without external support scoring better for the development of long-term partnerships. Secondly, the partnership projects that were designed to address intrinsic interests in both the German and Indian partner cities tend to score higher for achieving partnership mutuality.

Based on the findings of the index "Exchange Strategy" hypothesis 1 is thus replaced by the following two propositions:

Proposition 2: The involvement of external partnership moderators appears to be beneficial for the implementation of large-scale projects as part of transnational urban partnerships, but it tends to hinder the long-term development of these partnerships.

Proposition 3: Transnational urban partnership projects are more likely to realise partnership mutuality if from the outset they are designed to address the intrinsic interests of both partner cities.

6.3.2 Linkages between the Partnership Project and State Institutions

Hypothesis 2: The more a transnational urban partnership project is institutionalised into the state system, the more likely the project is to succeed. The second hypothesis argues that urban partnership projects benefit from being closely institutionalised into the state system. Table 28 highlights the results of Index 2 that measures the degree of state institutionalisation of the four urban partnership projects analysed.

	Success Index (total 14)	I1: Exchange Strategy (total 14)	**I2: State Institutionalisation (total 12)**	I3: Partnership Entrepreneur (total 12)	I4: Social Capital (total 14)
DEWATS	6	7.5	*2.5*	10	8.5
Tramway	4	7.5	*3.5*	10	6.5
W2E (tentative)	7	11	*9*	0	3
LRMCN	7.5	6	*7*	3.5	2.5

Table 28: Results of the Index System Scoring (Index 2 "State Institutionalisation" highlighted)

The total index scores vary significantly, with the four partnership projects reaching between 2.5 and 9 points in this index, out of a total 12 points possible. The scores thereby confirm that the degree of institutionalisation in the state system of the four urban partnership projects differs considerably. This is in accordance with the case selection rationale (see chapter 4.5); the two projects from the non-state city cooperation Pune-Bremen (the primarily NGO-run project DEWATS and the corporate actor-oriented tramway project) have relatively weak links with state institutions, the GIZ project W2E is strongly institutionalised in the state system and the city network project LRMCN is partially embedded in state institutions.

The scores for the four transnational urban partnership initiatives show a positive correlation between the total scores of the index "State Institutionalisation" and the total scores of the success index. The two projects that are more institutionalised in the state system (W2E and LRMCN) score slightly better in the success index than the two Pune-Bremen partnership projects (DEWATS and tramway) that have fewer links to state institutions. The correlation between the total index scores is however weak.

Significant correlations can only be identified between the results of the "State Institutionalisation" index and individual success indicators. In fact, the results of the "State Institutionalisation" index offer more insights on the above identified negative correlation between the two success indicators measuring macro-level project implementation and the long-term development of city partnerships.

In all four projects analysed, the scores for the "State Institutionalisation" index are positively correlated with the scores for the success indicator "Project Implementation" and negatively correlated with the scores for the success indicator "Long-term Partnership".

The positive correlation between "State Institutionalisation" and "Project Implementation" points to the conclusion that the closer transnational urban partnerships are linked to state institutions, the higher the chances are they will realise large-scale partnership projects; whereas those projects with weaker links to state institutions are more likely to face difficulties with macro-level implementation.

The latter is demonstrated by the two projects that fall under the Pune-Bremen partnership (DEWATS and tramway). Both initiatives only worked well on a micro level where they were largely coordinated and executed by local non-state actors (the NGO pilot projects in the DEWATS initiative and the DPR preparation in the tramway project). But these projects reached their limits when they required state support for their city-wide implementation. In the DEWATS initiative local state approval and resources were only provided for one pilot plant, and the wider dissemination of the DEWATS technology failed due to the lack of political and financial support from Pune's city administration. The tramway transfer project started with a well-functioning collaboration with private consultants in the preparation of the DPR. But the project implementation also required local, state and national government support which it did not receive and therefore the execution of the tramway transfer could not be realised.

The results for the individual indicators that make up the "State Institutionalisation" index (see table 29) specify why close linkages with state institutions tend to facilitate the realisation of larger-scale partnership projects. In all four partnership initiatives the implementation of larger scale projects required formal approval from local state institutions (indicator 1) and three initiatives (tramway, W2E and LRMCN) also depended on formal approval from the higher policy levels (indicator 2). Those initiatives that were primarily driven and implemented by local non-state actors (DEWATS and tramway) failed to receive formal state approval for the large-scale implementation of their projects. The W2E and LRMCN initiatives that were managed by external project coordinators (W2E and LRMCN) also initially struggled to receive the required state approval, particularly from the local government, but eventually they obtained the necessary approval and could realise the projects. It did not therefore matter whether the external coordinator itself was a state body, like the GIZ, or a semi-public organisation, like ICLEI (indicator 5). But it was however crucial that both the GIZ and ICLEI had good access to local and higher policy level decision makers and institutions. Both organisations also had the capacity to build up a continued presence in the Indian cities of Nashik and Nagpur which was required to sustain political ties despite the frequent changes in political leaders in these cities.

I2 Indicators	Formal Approval by State Institutions (Local)	Formal Approval by State Institutions ((Sub) National)	State Human Resources	State Financial Resources	State Project Coordinator	Commitment of local State Leaders	**I2: State Institutionalisation (total 12)**
DEWATS	+--	--	+-	+--	--	+--	2.5
Tramway	+-	--	+--	+--	+--	+-	3.5
W2E (tentative)	+-	++	++-	++-	++	+-	9
LRMCN	+-	+-	+-	++	+-	+-	7

Table 29: Index 2 "State Institutionalisation": Results of the Indicator Scoring

The analysis of the four partnership initiatives indicates moreover that the commitment and leadership of local state actors is beneficial or even a prerequisite for the realisation of large-scale projects in transnational urban partnerships (indicator 6). But all initiatives also reveal that the frequent change of political leaders who are often elected or appointed for only one or two year terms in Indian cities, is a special challenge and burden for transnational urban partnership projects. Partnership actors have to invest considerable resources in order to convince new political leaders of the projects' benefits, which is particularly difficult when partnership projects are managed and driven from Germany.

As for the availability of human and financial resources (indicators 3 and 4), the implementation of the W2E and LRMCN projects were clearly facilitated by the considerable shares of project funding provided by the German and Indian national governments. However, access to state funds is not necessarily a prerequisite for transnational urban partnership projects. This is demonstrated by the fact that both the tramway and W2E projects were designed as public-private-partnerships with private contractors running the projects. According to the different scenarios outlined in their business plans, both projects could still be profitable even if they did not receive any state funding.

While close links to state institutions thus seem to positively influence the realisation of macro-level projects, it does not appear to foster the long-term development of city partnerships beyond single projects. The identified negative correlation between the four cases' scores in the "State Institutionalisation" index and the "Long-term Partnership" indicator suggests that linking the partnerships with state institutions may even adversely affect the sustainability of partnerships. This conclusion is most strongly supported by the DEWATS and tramway initiatives as part of the Bremen-Pune cooperation. In contrast to the two other partnerships analysed in this study, the partnership between Pune and Bremen continued after the completion of the DEWATS and tramway initiatives. However due to the failure to involve state actors in these projects the Pune-Bremen partnership actors decided only to continue their cooperation on a non-state level.

The LRMCN and W2E initiatives have been from the outset designed as ad hoc and temporary partnerships. Both projects are driven and dependent on their external partnership moderators, the GIZ and ICLEI, who prioritised concrete project development with governmental actors in the Indian cities of Nagpur and Nashik. The setting up of long-term city partnerships beyond these individual projects was of minor relevance.

The comparison of the index results suggests that hypothesis 2 needs to be refined. In the four urban partnership initiatives analysed in this study only large-scale projects clearly benefited from close linkages with state institutions. This conclusion is not valid for examples of micro-level project collaboration which were successfully realised both with and without the support of state institutions. Furthermore, a negative correlation between state institutionalisation and the long-term development of city partnerships was identified in the cases analysed.

Hypothesis 2 in this study has therefore been modified resulting in the following proposition:

Proposition 4: Large-scale projects in transnational urban partnerships tend to rely on close linkages with state institutions, but the reliance on state institutions can also hinder the long-term development of these partnerships.

6.3.3 Partnership Entrepreneurs

Hypothesis 3: A transnational urban partnership project is more likely to succeed if it is driven by engaged, persuasive and well-networked partnership entrepreneurs. The third hypothesis suggests that the potential for realising urban partnership projects is greater if individuals from both partner cities take personal ownership of the initiation and implementation of the projects. Index 3 measures the involvement and engagement of partnership entrepreneurs in the four partnership projects analysed. The scores are highlighted in table 30.

	Success Index (total 14)	I1: Exchange Strategy (total 14)	I2: State Institutionalisation (total 12)	**I3: Partnership Entrepreneur (total 12)**	I4: Social Capital (total 14)
DEWATS	6	7.5	2.5	**10**	8.5
Tramway	4	7.5	3.5	**10**	6.5
W2E (tentative)	7	11	9	**0**	3
LRMCN	7.5	6	7	**3.5**	2.5

Table 30: Results of the Index System Scoring (Index 3 "Partnership Entrepreneur" highlighted)

The index scores differ substantially. They reveal that partnership entrepreneurs show widely varying degrees of involvement in the four projects analysed, with scores ranging from 0 in the W2E initiative to 10 in the DEWATS and tramway projects, out of a total of 12 points possible in this index. The Pune-Bremen DEWATS and tramway partnership projects achieve a particularly high score in this index. Local PEs were strongly involved in both initiatives; the projects were in fact co-initiated and driven throughout their duration by the local individuals Hilliges from Bremen and Mahajani (DEWATS and tramway) and Gujar (DEWATS) from Pune. The LRMCN gains a much lower score (3.5), as only one PE, the member of Freiburg's city administration, was involved for a short time at the beginning of the initiative. The W2E scores zero in this index as the initiative was not driven by local PEs from Nashik or Hamburg at any point.

A comparison of the index results shows a weak negative correlation between the scores of the "Partnership Entrepreneur" index and the success index; those projects in which PEs played a substantial role score slightly lower in the success index.

Closer analysis of the correlations between the "Partnership Entrepreneur" index scores and the scores for the individual success indicators once again reveals more significant results. In fact, the "Partnership Entrepreneur" index also correlates with the "Project Implementation" and "Long-term Partnership" success indicators. In contrast to the "State Institutionalisation" index and the "External Moderation" indicator (see above), the "Partnership Entrepreneur" index is, however, negatively correlated with the "Project Implementation" indicator and positively correlated with the "Long-term Partnership" indicators. The partnership projects that are driven by local PEs (DEWATS and tramway) gain very low scores for large-scale and city-wide implementation, but they score far higher for the continuation of the partnerships after the project's completion compared with the projects that feature only limited involvement of PEs (W2E and LRMCN) (see table 23).

The DEWATS and tramway initiatives thereby reflect the partnership between Pune and Bremen more generally. Individuals such as Hilliges, Mahajani and Gujar have been crucial in keeping the city cooperation running for several decades, but the focus of the partnership has been primarily on realising micro-level partnership projects between non-state actors. The separate scorings of partnership entrepreneurs from Bremen and Pune in the indicators measuring their access to relevant policy networks provide a possible explanation for this finding (see table 31). Only Hilliges from Bremen is comprehensively networked in and beyond the partner cities. Mahajani and Gujar from Pune had good personal relationships with Hilliges which have been instrumental in maintaining the exchange between Bremen and Pune for such a long period. However the pair's access to relevant policy networks was generally more limited and they were unable to mobilise sufficient support for a largescale implementation of partnership initiatives.

I3 Indicators: Access to Policy Networks			
	Internal Policy Network	External Policy Network	Partner City Network
Pune	+-	+-	+-
Bremen	++	++	++

Table 31: Indicators Access to Policy Networks: Scoring of the Indian and German Cities

The two partnership initiatives where PEs played a minor (LRMCN) or no role (W2E) realised or will likely realise large-scale projects, but are shown to be less successful at sustaining the partnership after these projects have been completed. In the LRMCN the member of Freiburg's city administration tried to maintain the direct exchange between Freiburg and Nagpur, but he remained unsuccessful in his attempts to contact his peers in Nagpur. Today, all direct contact between Freiburg and Nagpur has been lost and by the completion of the data collection for this study there were no plans for future cooperation. The W2E partnership was solely driven by the GIZ, rather than by local individuals from Nashik or Hamburg and a long-term development of the partnership between Hamburg Wasser and Nashik MC appears unlikely outside of the GIZ framework.

Consequently, hypothesis 3 must also be refined. According to the findings from the four cases analysed in this study, partnership entrepreneurs appear to play an instrumental role in the realisation of long-term cooperation. They can also play a positive role in the initiation and implementation of projects on a micro level but their capacities and influence tends to be more limited when it comes to realising macro-level projects. Hypothesis 3 is thus replaced by the following proposition:

Proposition 5: Engaged, persuasive and well-networked partnership entrepreneurs can play an instrumental role in initiating and maintaining transnational urban partnerships and driving micro-level partnership projects. However they appear to play a more minor role in the successful implementation of larger scale projects.

6.3.4 Partnership Social Capital

Hypothesis 4: The more social capital protagonists develop as part of the transnational urban partnership, the more likely the partnership project is to succeed. The fourth hypothesis in this study suggests that partnership actors from both cities must possess sufficient social capital in order to successfully realise urban partnership projects. Table 32 highlights the four partnership projects' scores in the index "Social Capital".

	Success Index (total 14)	I1: Exchange Strategy (total 14)	I2: State Institutionalisation (total 12)	I3: Partnership Entrepreneur (total 12)	**I4: Social Capital (total 14)**
DEWATS	6	7.5	2.5	10	**8.5**
Tramway	4	7.5	3.5	10	**6.5**
W2E (tentative)	7	11	9	0	**3**
LRMCN	7.5	6	7	3.5	**2.5**

Table 32: Results of the Index System Scoring (Index 4 "Social Capital" highlighted)

The scores for the "Social Capital" index point to substantial differences between the four cases analysed with regard to the development of social capital in urban partnership initiatives. The scores range from 2.5 in the LRMCN project to 8.5 in the DEWATS project based on a scale of 0 to 14.

Similar to the "Partnership Entrepreneur" index, a weak negative correlation is evident between the total scorings for the "Social Capital" index and success index; the two initiatives that receive low scores in the index measuring the partnership social capital (LRMCN and W2E) gain higher scores in the success index, whereas the two projects conducted as part of the Pune-Bremen city partnership (DEWATS and tramway) score higher in the "Social Capital" index but lower in the success index.

Again, a comparison of the individual indicators reveals more significant explanatory correlations and patterns (see table 33).

I4 Indicators	Partnership Experiences	Trust	Equality	Exposure to International Exchange	Project Communication	Informal Leadership Networks	Inclusion & Participation	I4: Social Capital (total 14)
DEWATS	++	++-	+-	++	+-	+-	+-	8.5
Tramway	++	+-	+-	+-	+-	+--	--	6.5
W2E (tentative)	--	+-	+-	+--	--	--	+--	3
LRMCN	--	--	--	+-	--	+-	+--	2.5

Table 33: Index 3 "Partnership Social Capital": Results of the Indicator Scoring

The DEWATS initiative is the only project analysed that fulfils all the indicators in the "Social Capital" index either fully or at least partly. However despite the existence of a relatively high level of partnership social capital only micro-level success was achieved in the form of DEWATS demonstration plants in Pune, whereas the city-wide dissemination of the DEWATS technology failed. The tramway initiative gains a slightly lower score in partnership social capital and similarly failed to implement the project on a larger scale. The partnership between Pune and Bremen however continued after these two projects were completed.

Partnership social capital appears to have played a minor role in realising large-scale project implementation in the LRMCN and the W2E. Both initiatives gain low scores in the "Social Capital" index and do not fully meet any of the indicators in the "Social Capital" index. Still, both projects have or are likely to realise their implementation targets. The two city partnerships are however unlikely to continue after the completion of the projects.

Considering the individual indicators in "Social Capital" index, the divide between the Pune-Bremen projects and the two other projects, W2E and LRMCN, is particularly evident in the two indicators measuring experiences from prior partnership projects (indicator 1) and project communication (indicator 5). In terms of prior partnership experiences (indicator 1) the DEWATS and tramway initiatives built on extensive knowledge gained from the many previous projects that took place as part of the city partnership between Pune and Bremen. Several of the protagonists involved (Hilliges, Mahajani, Gujar) and organisations (BORDA, MAM, AWO, LAFEZ, PMC) had already collaborated with each other during previous partnership initiatives. Such existing relationships did not exist between the partner cities of Nashik and Hamburg in the W2E project and between Nagpur and Freiburg in the LRMCN initiative.

As for project communication indicator (indicator 5), actors in the DEWATS and tramway initiatives interacted directly with one another and without the aid of external moderators. The communication worked well overall and initial language barriers were overcome. In the W2E (Nashik-Hamburg) and LRMCN (Nagpur-Freiburg) initiatives the project communication between the partner cities depended fully on the external moderators, the GIZ and ICLEI. In the case of the LRMCN, more direct exchange was in fact hindered by communication problems as the project partner from Freiburg was unable to contact his peers from Nagpur by email.

The partnerships' scores also differ in the indicator measuring trust between the protagonists (indicator 2), a crucial component of partnership social capital. At the beginning of both the DEWATS and tramway initiatives high levels of trust existed between the actors involved. In the DEWATS projects a trusting atmosphere between the non-state actors involved continued throughout the duration of the projects, whereas trust in state actors decreased. In the W2E and LRMCN initiatives personal, trusting relations between the German and Indian city actors were not developed due to the lack of direct exchange between these individuals. In both initiatives the city actors did however show high trust in their respective external partnership moderators, the GIZ and ICLEI.

In the remaining indicators measuring partnership social capital the divide between the DEWATS and tramway initiatives and the W2E and LRMCN projects is less evident. None of the four cases scores highly for perceived equality between the German and Indian project partners. The Pune-Bremen partnership projects attempted to realise more equal partnerships (indicator 3) in the DEWATS and tramway projects by trying to engage the local (DEWATS) and sub-national state governments (tramway) from India as projects funders. But in both cases the Indian funding remained limited and the German actors remained the main drivers of the projects. Also the W2E project has been continuously driven by the GIZ and the flow of information has been largely one-directional, namely from Hamburg Wasser to Nashik. The LRMCN partnership was from the outset designed as a one-sided partnership with Freiburg providing Nagpur with consultancy on the development of a local clean energy strategy. With regard to the partnership actors' intercultural experiences, many of the directly involved protagonists in the four partnership initiatives had already worked in international projects before (indicator 4). However, with the exception of the DEWATS project, the majority of the actors in the other three initiatives had never worked in or together with actors from the respective partner country so their experience of Indian-German collaboration was limited. The levels of stakeholder involvement (indicator 6) and inclusion (indicator 7) were generally low in all four cases. Only in Nagpur in the LRMCN project were local informal leadership groups and local citizens involved in the partnership activities to a greater extent.

Summing up the findings on the development of partnership social capital in the four cases leads to the conclusion that the fourth hypothesis in this study also needs to be refined. The W2E and LRMCN projects demonstrate that social capital among the city actors involved is not necessarily a prerequisite for realising projects in transnational urban partnerships, especially not for large-scale projects. All four cases however indicate that partnership social capital appears to be a precondition for developing long-term cooperation between cities. In particular experiences from prior partnership work, partnership protagonists' intercultural competences, plus personal, trusting relations between both partner cities tend to play an important role in sustaining partnerships once individual projects have been completed. Hypothesis 4 has thus been modified into the following proposition:

Proposition 6: The development of social capital in transnational urban partnerships is a key prerequisite for realising long-term cooperation beyond single projects. However partnership social capital appears to play a minor role when it comes to the implementation of large-scale projects.

6.4 Synopsis of Comparative Analysis

The comparative analysis of the index results has led to this study's four hypotheses being modified, resulting in six new propositions regarding the conditions for success and failure in transnational urban partnerships.

With the exception of proposition 3 all other propositions speak to the identified negative correlation between the two success Indicators measuring macro-level project implementation and long-term development of urban partnerships. The empirical basis of these propositions is summarised once again in table 34.

The overview visualises the negative correlation between the two success indicators as well as their positive and negative correlations with the indicator "External Moderation" and the three indexes "State Institutionalisation", "Partnership Entrepreneur" and "Social Capital". Table 34 indicates that three additional propositions can be drawn. Proposition 7 sums up the propositions 2, 4, 5 and 6 with regard to the success conditions for large-scale partnership projects and the long-term development of partnerships:

Index	Success Index		I1: Exchange Strategy	I2: State Institutionalisation (total 12)	I3: Partnership Entrepreneur (total 12)	I4: Social Capital (total 16)
Indicator	Implementation Achievements	Post-Project Sustainability	External Moderation			
DEWATS	+-	+-	--	2.5	10	8.5
Tramway	--	+-	--	3.5	10	6.5
W2E (tentative)	++	--	++	9	0	3
LRMCN	++	--	++	7	3.5	2.5

Table 34: Results of the Index System Scoring (Selected Indexes and Indicators)

Proposition 7: Whereas external partnership moderation and state institutionalisation appear to be crucial success factors for realising large-scale projects, the involvement of PEs and partnership social capital seem to be instrumental for the long-term development of city partnerships.

Furthermore, table 34 demonstrates that in the four cases analysed negative correlations exist between key independent variables. In fact, the index results point to certain trade-offs, or dilemmas, faced by the four urban partnership projects. The partnership projects either gain high scores for the involvement of external moderators and state institutionalisation and low scores for the involvement of PEs and the development of social capital (W2E and LRMCN) or vice versa. This leads to the following proposition:

Proposition 8: There appears to be a trade-off between the involvement of partnership entrepreneurs and partnership social capital development versus the involvement of external partnership moderators and the linking of partnerships with the state system.

Together, propositions 1, 7 and 8 point to the conclusion that transnational urban partnerships tend to be either set up in a bottom-up approach, based on social capital, driven by partnership entrepreneurs and leading to long-term collaboration on a micro level; or they are induced in a top-down manner, driven by external part-

6.4 Synopsis of Comparative Analysis

nership moderators with close links to state institutions, and a focus on single project implementation rather than long-term partnerships. These different approaches towards designing urban partnerships are summarised in the ninth proposition:

Proposition 9: There appear to be two conflicting approaches to transnational urban partnership development; partnership projects that are set up in a bottom-up manner tend to facilitate long-term collaboration but only on a micro level; whereas top-down induced partnership projects tend to facilitate macro-level project implementation, but the lifespan of the partnership tends to be shorter.

The validity of the partnership dilemmas proposed and potential ways to approach them will be elaborated in more depth in the following chapter by linking the findings to the existing literature.

In the next chapter the validity of proposition 3 will be also be discussed in more detail; drawing a connection between the mobilisation of intrinsic interests of both partner cities and the realisation of partnership mutuality (see table 35).

Index	Success Index	I1: Exchange Strategy
Indicator	Mutuality	Intrinsic Interests
DEWATS	--	+--
Tramway	++	++-
W2E (tentative)	++	++
LRMCN	--	+-

Table 35: Results of the Index System Scoring (Selected Indicators)

7 Discussion and Conclusions

The case study findings demonstrate that transnational urban South-North cooperation on low carbon transitions is possible and is emerging in various institutional forms. In addition to more traditional twinning arrangements, cities also connect in ad hoc and project-oriented partnerships, which can be driven and moderated by external moderators such as the GIZ or ICLEI. The data however points to key challenges that remain in transnational urban partnerships. The first major challenge faced by such collaboration is widening the scope of the cooperation beyond micro-level cooperation and demonstration projects. A second key challenge is ensuring the post-project sustainability of transnational urban partnerships. A third challenge is achieving mutuality, and the perception of it, in relationships between partners from the Global South and North.

This chapter explores these challenges in greater depth, linking them to related debates in the literature on the concepts of multi-level governance coordination, mutuality and co-benefits, to discuss the broader validity of the identified challenges. Based on these discussions, lessons will be drawn on how to design and implement transnational urban partnerships in a more efficient and effective manner, to mobilise the still largely untapped potential of urban knowledge exchange, learning and cooperation in addressing global warming. The chapter concludes by outlining academic implications and recommendations for future research.

7.1 Key Factors Determining the Success and Scope of Urban North-South Cooperation

7.1.1 The Potential of Multi-Level Governance Coordination to Address the Gap between Bottom-up and Top-Down Approaches

It can be argued that the tensions between bottom-up versus top-down approaches in the design of transnational urban partnerships that have been identified in this study reflect the general lack of multi-level coordination in both global climate governance and international development cooperation.

Since its initiation at the Rio Conference in 1992 global climate governance has been increasingly characterised by a multiplicity of governance networks in various institutional forms. These networks can be driven by both state and non-state actors and institutions, or combinations thereof, at and beyond several policy levels. In addition to the formal UNFCCC climate negotiations between nation states numerous transnational and subnational initiatives have formed. Some of these initiatives aim to support and complement the formal international negotiations, while others follow their own agenda and experiment with alternative approaches to tackling global warming.

The increasing diversity within global climate governance offers much potential for innovation and the diffusion of best practises, as it fosters competition between the various bodies to come up with the most effective approaches to climate-friendly development. City-level initiatives in particular have often been highlighted for their function as first-movers and laboratories for experimentation (Alber and Kern, 2008; Schreurs, 2008; Anguelovski and Carmin, 2011; Cástan Broto and Bulkeley, 2013).

At the same time the growing number of subnational and transnational climate networks also adds to the complexity and fragmentation of global climate governance. The emergence and consequences of an increasingly fragmented system of global environmental and climate governance have been documented and discussed in several research projects and publications (Bäckstrand, 2008; Biermann et al., 2012; Corfee-Morlot et al., 2009; Isailovic, Widerberg and Pattberg, 2013; Pattberg et al., 2014; Prins et al., 2010). There is a risk that without better coordination and integration of national top-down approaches and less formalised bottom-up initiatives inefficiencies, redundancies, or even contradictory developments may hamper global climate protection efforts. According to Bäckstrand (2008), multi-level coordination and the adequate integration and accounting of non-nation state climate action are among contemporary climate governance's key challenges.

Corfee-Morlot et al. (2009) emphasise that cities' climate change initiatives are particularly prone to remaining isolated from activities in other cities as well as from national climate strategies. The authors highlight that more governance coordination is required for the "narrowing or closing of the policy 'gaps' between levels of government via the adoption of tools for vertical and horizontal cooperation" (ibid., 7-8). Corfee-Morlot et al. call for a strengthening of vertical cooperation, stating that national governments need to work more closely with local and regional governments to implement national climate programs and they must provide subnational governments with adequate financial support and institutional frameworks. They also argue that horizontal cooperation between state and non-state actors needs to be extended to improve the sectoral integration of climate change-related policy fields, enhance stakeholder involvement and foster learning and the diffusion of experiences (ibid.). Similarly, Alber and Kern (2008) point to the need to improve the governance coordination of cities' climate activities. They explain that in most countries there exist no direct links between national GHG emission reduction goals and their implementation at the regional and local level. Cities' climate protection engagement remains mostly a voluntary task which leads many cities to show passivity towards climate action or limit their efforts to cost-efficient measures in energy efficiency projects. Schreurs (2010) highlights that as a result of the lack of horizontal and vertical governance coordination between state and non-state actors, urban knowledge exchange in urban climate response often also remains ineffective and suboptimal.

Fragmentation and incoherence is also a well-known challenge in the policy field of international development cooperation. Stockmann, Menzel and Nuscheler (2010) explain that poor coordination between actors and institutions can greatly hamper the implementation of programs and projects. They point to how weak horizontal coordination between development cooperation strategies of different donor countries results in unnecessary transaction costs for donor countries and the overburdening of recipient countries. Stockmann et al. also identify a lack of coordination between governmental departments in many countries, and between and among state and non-state actors conducting development cooperation. It is notable that the authors consider institutional fragmentation as primarily a barrier for the realisation of state-led development cooperation. They see fragmentation among non-state actor-led development cooperation projects as less problematic, as it represents a pluralism of societal interests, worldviews and attitudes, plus adds to the diversity of civil society institutions within the recipient countries (ibid., 455-456). Stockmann et al. however acknowledge that there is still too little comparative research available on the effectiveness of state versus non-state development cooperation, a finding that is also supported by Riddell (2007).

As for the vertical coordination of individual initiatives and top-down strategies in development cooperation, practical experiences point to challenges similar to the dilemma between bottom-up versus top-down development of transnational urban climate cooperation outlined above. A project report by the European Commission-funded "Platform of local and regional authorities for development" (PLATFORMA) (2011) concludes that there is a "fine balance to strike between the need to coordinate activities and avoid duplication, whilst recognizing the importance of individual initiatives by local authorities which can lead to positive long-term relationships and benefits" (ibid., 10).

In Germany and India the contested allocation of cities' legal competences within the federal political systems further adds to the need for multi-level governance coordination. In Germany, the debate about the legal rights of cities to conduct international cooperation is still ongoing and undecided. Nitschke et al. (2009) explain that it is not clear whether cities' constitutionally granted rights to municipal self-government (Art. 28 Basic Constitutional Law) covers the mandate to establish international relations, or whether the competence for foreign diplomacy lies solely with the national level (Art. 32 Basic Constitutional Law). In order to avoid legal conflicts, Nitschke et al. recommend that city governments align their international cooperation activities closely with the national government's development cooperation framework. They also highlight horizontal integration of non-state actors as a key success factor for urban development cooperation as most existing initiatives are driven and sustained by civil society networks.

Multi-level coordination of local initiatives is also crucial in India, as Indian cities rely greatly on both the political and financial support of higher governmental policy levels. Bhagat (2005) states that in India's phase of liberalisation reforms in the early 1990s greater emphasis was put on the decentralisation of authority to the urban level to foster self-reliant, competitive and efficient governance by urban local bodies. In 1992 the Indian parliament adopted the 73rd and 74th Amendments of the Constitution, granting local governments a constitutional status in India's federal system and obliging them to prepare their own development plans and generate their own resources (ibid.). The constitutional amendments have however not been comprehensively implemented and a significant decentralisation of authority to the city level has yet to be realised. Urban development policies are still strongly shaped by state governments and cities' political, institutional and financial capacities remain limited (Sharma and Tomar 2010).

The four cases in this study demonstrate the difficulties and tensions arising from efforts to harmonise bottom-up initiatives with top-down coordination in transnational urban cooperation. The four Indian-German partnership projects analysed illustrate that realising and sustaining urban cooperation, particularly between cities from different global contexts, requires the partnership actors to have networking capacities to engage decision makers and stakeholders both locally and at higher policy levels. The case study findings thereby support Bäckstrand's (2008, 75-76) conclusion that most transnational "climate partnerships operate in the 'shadow of hierarchy'" as in all four partnership initiatives a (positive or negative) correlation could be identified between the projects' linkages with the multi-tiered state system and their successful large-scale realisation.

These results indicate that improving multi-level governance coordination is also key to addressing the challenges posed by bottom-up versus top-down approaches to the development of transnational urban North-South partnerships. The case study findings suggest that a crucial step forward in the facilitating of multi-level governance coordination could be to focus on better connecting partnership entrepreneurs and external moderators, as they tend to have distinct yet complementary roles and capacities. Partnership entrepreneurs can be important actors in helping to integrate local non-state stakeholders in transnational urban partnership activities. They possess in-depth insights into the local context and they often know and have access to relevant local stakeholder groups. This is exemplified by partnership entrepreneur Hilliges from the Pune-Bremen cooperation who was instrumental in establishing a partnership network, consisting mainly of non-state actors, which has driven the partnership activities for more than 30 years. External partnership moderators, such as ICLEI in the Nagpur-Freiburg partnership and the GIZ in the Nashik-Hamburg cooperation, tend to be more important actors in facilitating the vertical integration of transnational urban partnerships in the multi-tiered governmental system. Both initiatives largely depended on ICLEI and the GIZ, respectively, for securing the necessary local, state and national government approval. The partnership moderators were also decisive in securing state funding for the partnership projects.

As partnership entrepreneurs appear to be important facilitators of horizontal governance coordination whereas external moderators tend to be more important for vertical governance coordination, it would be advantageous to include both actors in transnational urban partnership projects and provide platforms for communication between them. This approach however may not be entirely straightforward, as the cases analysed in this study demonstrate. In none of the four German-Indian urban low carbon initiatives did partnership entrepreneurs and external moderators effectively join forces. In fact, the four initiatives were either led by partnership entrepreneurs without involvement of external moderators (the DEWATS and tramway projects as part of the Pune-Bremen cooperation); or they were driven by external moderators without or with limited involvement of partnership entrepreneurs (Nashik-Hamburg and Nagpur-Freiburg).

This study concludes therefore that the development of local partnership centres or meeting points could help institutionalise communication and coordination between all actors involved or affected by partnership projects, including partnership entrepreneurs and external moderators. This corresponds with the statements by actors interviewed for this study highlighting the need to focus more on establishing "structures" to increase the efficiency and impact of future transnational urban cooperation (Interview with Coimbatore Municipal Corporation, 13-12-2012; Interview with NGO BORDA, 11-09-2014). According to these actors, a crucial strategy would be the establishing of local offices in the partner cities which would serve as permanent gathering points and spaces for the organising, conducting and documenting of partnership work. Local partnership centres would facilitate coordination and communication not only between the partner cities, but also between the local actors involved in partnership activities. A permanent partnership office could also be an important prerequisite for building and sustaining know-how and capacity in partnerships which may also help reduce the dependency on the support of local political leaders who are subject to electoral cycles.

In fact, all the partnerships analysed in this study have some experience of setting up local offices in the Indian partner cities. ICLEI established the Renewable Energy and Energy Efficiency Resource Centre in Nagpur to manage the implementation of the LRMCN project. The GIZ coordinates the implementation of the W2E project from its "Environmental Cell" in Nashik. The GIZ also aims to develop a training centre in Nashik to host workshops for city managers from other Indian cities and facilitate lesson-learning from the W2E and other waste management projects in Nashik. Like the overall partnership initiatives these two offices however remain strongly dependent on the engagement of ICLEI and the GIZ and are unlikely to survive without these external partnership moderators. Also in the Pune-Bremen cooperation the partnership actors set up a local office in Pune, the International Office Agenda 21, to strengthen and institutionalise partnership activities but a lack of financial and personnel meant that it never developed into a thriving partnership centre.

Two approaches that demonstrate how local knowledge centres can effectively bring together Indian and European partners in low carbon and sustainable development are the Consortium for DEWATS Dissemination (CDD) Society in Bangalore and the Cultural Heritage and Management Venture lab in Ahmedabad. Based on the experiences from its DEWATS projects under the Pune-Bremen partnership and other pilot projects in India, the NGO BORDA from Bremen established the CDD Society[56] in Bangalore, developing it into a leading competency centre on DEWATS, community-based sanitation and city sanitation planning in Asia. The Cultural Heritage and Management Venture lab in Ahmedabad[57] was set up as part of a partnership between the cities of Ahmedabad (India) and Valladolid (Spain). Funded by the European Union, local experts from both cities and their universities together with local citizens from Ahmedabad developed a strategy and knowledge centre focusing on how Ahmedabad can draw Valladolid's experiences of heritage management and preservation. The project addresses several aspects of local sustainable development by simultaneously preserving and promoting the cultural heritage of traditional housing, which is eco-friendly and fosters social unity and neighbourhood development. The initiative also focuses on developing innovative approaches to the linking of heritage preservation and economic values, for example through tourism development.

---

56 http://www.cddindia.org/ (19-02-2016)
57 www.ahmedabadheritagecluster.com/ (19-02-2016)

In addition to proposing the need for investment in local partnership centres, this study also concludes that a key strategy for the successful implementation of transnational urban partnership projects is establishing and strengthening links with existing networks that foster urban exchange and learning. Such networks have emerged at all governance levels, many of them targeted at improving multi-level governance integration, transparency and the global transferability of urban climate activities.

Examples of such networks are transnational municipal networks (TMNs). The Nagpur-Freiburg case study shows that cities can benefit greatly from engagement with TMNs. ICLEI played a key role in establishing contact with relevant national and local decision makers and securing funding for the LRMCN project implementation in Nagpur. Plus ICLEI also provided Nagpur with technical, administrative and political know-how on low carbon transformation based on Freiburg and other cities' experiences. While ICLEI was instrumental in organising funds and providing contacts, the LRMCN case study also supports the finding from other studies that TMNs still struggle to facilitate direct knowledge exchange and long-term cooperation between their member cities (Betsill and Bulkeley, 2004; Medearis and Dolowitz, 2013). An important step in addressing this gap was the recent Compact of Mayors initiative[58], a joint coalition of the three TMNs ICLEI, C40 and United Cities and Local Governments (UCLG) which aims to develop a common emissions inventory method. The Compact of Mayors was officially launched by UN Secretary General Ban Ki-moon and his Special Envoy for Cities and Climate Change, Michael R. Bloomberg, at the UN climate summit in New York in September 2014. The initiative intends to establish "robust and transparent data collection standards" and "common reporting processes for local climate action that allow for consistent and reliable assessment", in order to quantify cities' climate mitigation commitments and their impact, plus facilitate lesson-drawing and cooperation (C40, UCLG, ICLEI, 2014). Participating cities are requested to publicly disclose climate mitigation targets and annually report on their progress using a standardised method, the Global Protocol for Community-Scale Greenhouse Gas Emission Inventories (GPC). The GPC method was developed by C40, ICLEI and the World Resources Institute (WRE) and made available in the context of the COP 20 in Lima in December 2014. If ICLEI, C40 and UCLG realise their target of making the GPC "the new globally recognized standard for community scale emissions reporting" (C40, UCLG, ICLEI, 2014), a major barrier to direct urban climate policy learning and transfer, i.e. the lack of standardised emissions inventory methodologies, could be removed. The first steps have been taken with 436 cities representing 378 million inhabitants and 5.2% of the world's

---

58   http://www.compactofmayors.org/ (19-02-2016)

population having formally committed to the Compact of Mayors and the GPC method as of December 2015.[59]

Cities that are interested in strengthening their international engagement in climate partnerships should also investigate whether they can make better use of support programmes provided by national and supranational governments. In 2007 the European Commission introduced a funding programme to strengthen local governments' and non-state actors' involvement in development cooperation. Under the framework of the European Development Cooperation Instrument (DCI) the "Non-State Actors – Local Authorities" programme provided 30 million Euros to local government development projects annually between 2008 and 2013.[60] The German government has also set up platforms and funding programs to support transnational urban partnerships, such as the "50 Municipal Climate Partnerships until 2015" project run by the Service Agency Communities in One World and the LAG 21 NRW on behalf of the Federal Ministry for Economic Cooperation and Development. The project facilitates and supports cooperation between German municipalities and municipalities in the Global South in the areas of climate mitigation and adaptation (ENGAGEMENT GLOBAL – Service for Development Initiatives, Service Agency Communities in One World, 2014). As of 2015 the project had provided a total of 43 partnerships between German and African and Latin American municipalities with financial support as well as technological and methodological advice. In summer 2015 the project was also extended to German-Asian urban partnerships.[61]

The Government of India's Smart City Mission may also yield opportunities for greater financial and institutional support for transnational urban cooperation. The Smart City Mission is an extensive program introduced by Prime Minister Modi in April 2015 which aims to boost "sustainable and inclusive development" in 100 Indian model "Smart Cities" and replicate their experiences in other Indian cities (Government of India. Ministry of Urban Development, 2015, 4). The Indian government is providing a total of 48,000 crores Indian Rupees (about 6.6 billion Euros) over five years to the 100 selected cities (i.e. on average 100 crore Indian Rupees, about 13.8 million Euros, per city per year), with the expectation that the respective state and city authorities match these funds. Although the Mission proposes several focus areas with close links to climate mitigation, such as renewable energy deployment, energy efficient housing and smart energy, transport and water management, it recognises the city's differing demands and contextual conditions and it leaves it to the cities themselves to define their individual focus areas

---

59   http://www.compactofmayors.org/ (19-02-2016)
60   http://www.platforma-dev.eu/page.php?lg=en&page_id=16&parent_id=2 (19-02-2016)
61   http://www.service-eine-welt.de/klimapartnerschaften/klimapartnerschaften-neue-kommunale-klimapartnerschaften-mit-asien.html (19-02-2016)

7.1 Key Factors Determining the Success and Scope of Urban North-South Cooperation

within "smart" city development. The Mission also places great emphasis on improving citizen participation and poverty reduction. In the drawing up of the Smart City Proposals (SCPs) the Government of India explicitly welcomes technical assistance from domestic, foreign and international organisations, mentioning amongst others the German government-owned development bank, the KfW (Government of India. Ministry of Urban Development, 2015).

The topic of "Smart Cities" may indeed become a focal area for future cooperation activities involving German and Indian cities, as evident in the 10th Asian-Pacific-Weeks on "Smart Cities" hosted by the state of Berlin in May 2015. At the event experts from Germany and Asian countries discussed innovative and intelligent solutions for urban water and energy supply, transport and sustainable city development.[62]

Considering the important role that subnational states often play in urban development, transnational city partnerships should also explore whether they can connect with existing state partnerships. The momentum of such partnerships is growing internationally, as the example of the U.S. – China Climate Leaders Declaration demonstrates. In this declaration several states, provinces and cities from the U.S. (amongst others California, Connecticut, Atlanta, Boston, Los Angeles and Seattle) and China (amongst others Beijing, Sichuan, Hainan, Shenzhen and Guangzhou) agreed to share experiences of climate mitigation and climate resilience strategies and strengthen their bilateral cooperation activities[63]. In Germany and India, two partnerships between Indian states and German Bundesländer are active in policy fields related to urban climate governance. Since 2007 the Indian state of Karnataka and the German Bundesland of Bavaria have cooperated on issues in the fields of sustainable agriculture and food. At a meeting of ministers in March 2015 the two states agreed to extend this cooperation to also address solar and bioenergy deployment.[64] Since January 2015, a new partnership has been established between the Indian state of Maharashtra and the German Bundesland of Baden-Württemberg, with urban infrastructure development as one of the major areas of cooperation.[65] As part of this state partnership two city partnerships are already starting to evolve. A delegation from Karlsruhe including Mayor Frank Mentrup visited Pune to promote cooperation between the two cities in the areas of clean and smart urban development as well as economic cooperation. Also the

---

62 http://apwberlin.de/en/apw2015/ (19-02-2016)
63 https://www.whitehouse.gov/sites/default/files/us_china_climate_leaders_declaration_9_14_15 _730pm_final.pdf (19-02-2016)
64 http://www.india.diplo.de/Vertretung/indien/en/02__Bangalore/00__start/bay__brunner.html (19-02-2016)
65 http://www.baden-wuerttemberg.de/de/service/presse/pressemitteilung/pid/friedrich-und-maharashtras-industrieminister-subhash-desai-unterzeichnen-laenderpartnerschaft/ (19-02-2016)

existing city partnership between Mumbai and Stuttgart aims to strengthen exchange and collaborative activities in the areas of urban planning as well as environmental and energy governance.[66]

In addition to state-led cooperation platforms transnational urban partnerships are also driven by non-state organisations and donors. An example is the Rockefeller Foundation that funds and organises several networks fostering transnational cooperation in the field of urban resilience to climate change. In 2008 the Rockefeller Foundation launched the Asian Cities Climate Change Resilience Network (ACCCRN) supporting ten cities in India, Vietnam, Indonesia and Thailand in setting up and sharing experiences on climate resilience strategies. In 2016 the ACCCRN plans to expand to include 50 new cities and two countries (Bangladesh and the Philippines).[67] The Rockefeller Foundation launched an additional, globally oriented urban climate resilience network, the 100 Resilient Cities Network, in 2013. According to the Foundation several hundred cities have applied to join the network, 67 cities have been pre-selected and by 2016 a total of 100 cities will have adopted the network's City Resilience Framework.[68]

The EuroIndia Centre is another NGO fostering climate cooperation between cities in India and Europe. Founded in 2001 by the Prime Ministers of France and India, Raymond Barre and Manmohan Singh, to foster subnational cooperation between Indian and European cities, regions and states, the EuroIndia Centre in 2006 adopted sustainable urban development as its major focus. The EuroIndia Centre has facilitated city partnership initiatives between Ahmedabad and the cities of La Rochelle (France), Valladolid (Spain), and Halle (Germany). In collaboration with European Business and Technology Centre (EBTC) the EuroIndia Centre also established an online query service to extend direct city exchange activities between European and Indian cities. On this online platform Indian and European cities can express their interest in cooperation in specific areas of urban development.[69]

Research projects are also a channel through which to foster communication and learning on low carbon transition between cities. One example is the research project "The Economics of Low Carbon Cities", led by Andy Gouldson from the University of Leeds. Oriented on the Stern (2007) Review on the Economics of Climate Change the project develops investment models which help cities and

---

66 http://www.india.diplo.de/Vertretung/indien/en/05__Mumbai/GermanArea/Visits__from__Germany__4__Delegations__Jan_2715__Seite.html (19-02-2016)
67 https://www.rockefellerfoundation.org/our-work/initiatives/asian-cities-climate-change-resilience-network/
68 https://www.rockefellerfoundation.org/our-work/initiatives/100-resilient-cities/ (19-02-2016)
69 http://www.the-euroindia-centre.org/content/default.aspx (19-02-2016)

metropolitan regions in Great Britain and countries of the Global South to transform in a profitable and climate-friendly manner. So far, investment models have been set up for Leeds, Birmingham, The Humber, Sheffield (UK), Kolkata (India), Lima-Callao (Peru), Palembang (Indonesia), Johor Bahru and Pasir Gudang (Malaysia) and Recife (Brazil). Green growth investment models for the Chinese cities of Beijing, Shanghai, Tianjin and Chongqing are currently under construction.[70]

7.1.2 Pursuing Intrinsic Interests and Co-benefits to Strengthen Mutuality in Urban North-South Partnerships

In addition to multi-level governance coordination this study demonstrates that another major challenge remains the lack of mutuality in transnational urban partnerships between cities from the Global North and South. Two of the four cases, the DEWATS and LRMCN projects, received particularly low scores for (perceived) mutuality. Both projects were implemented exclusively in the Indian cities, whereas the projects had no direct impact in the German partner cities. The two other partnerships, the tramway and the W2E initiatives, score higher in the indicator measuring partnership mutuality. Yet the implementation of these two projects is also only planned for the Indian partner cities. The perception of who benefits from the partnerships is less one-sided in these projects. The results of this study suggest that a major variable for the extent of mutuality reached in urban North-South partnerships is the degree to which both partner cities pursue their own intrinsic interests within the initiatives and thereby have a stake in the projects' successful realisation. This finding is supported by several studies which conclude that in order to realise greater mutuality and equality in urban partnerships, Northern and Southern partner cities need to be able to identify how they benefit from the cooperation and how they can communicate these benefits within their cities (Bontenbal and van Lindert, 2009; Johnson and Wilson, 2009; van Ewijk and Baud, 2009).

Although this conclusion appears to be straightforward, many urban North-South partnerships still struggle to design projects that simultaneously address intrinsic interests in both partner cities. In particular cities from the Global North often find it difficult to clearly identify and communicate how partnerships with cities from the Global South serve their own self-interests, as van Ewijk and Baud (2009, 218) explains: "In the literature on city-to-city partnerships, mutuality is considered an aim of most municipalities but 'benefits' for municipalities in the

---

70 http://www.climatesmartcities.org/ (19-02-2016)

North remain unclear in practice, and the notion of mutuality is generally not explicit." One potential reason for the lacking focus on the benefits for Northern cities is that partnership actors often think that practises and programs are generally more "advanced" in cities from the Global North and that these cities therefore have nothing to learn from the "less advanced" practises of their peers in the Global South. This view appears to represent a limited and primarily technology-oriented perspective on transnational urban cooperation. Several studies demonstrate that Northern cities can gain practical benefits from exchanging experiences from a range of areas with Southern cities. Examples include the exploration of new markets for local industries and businesses, innovative models of participatory governance, improving mutual understanding and relationships between established and migrant communities, plus the educational and personal development of the actors participating in the partnerships (Bontenbal, 2009; Bontenbal and van Lindert, 2008; Devers-Kanoglu, 2009; Johnson and Wilson, 2009; van Ewijk and Baud, 2009).

Another barrier to the realising of mutuality and equality in transnational urban partnerships is the fact that in most cases, the majority of the project funding is provided by actors or institutions from the Global North (Bontenbal and van Lindert, 2008; Johnson and Wilson, 2009). This can lead to power imbalances and unilateral dependencies within partnerships. Johnson and Wilson (2009, 211) call this challenge the "mutuality gap" in urban partnerships between cities from the Global North and South: "Differences from which partners can potentially learn may be hidden by differences of status and influence in the partnership, and such inequalities have the potential to undermine the incentive for engagement" (ibid.). Power imbalances resulting from one-sided partnership funding can also prevent open discussions and negotiations on the specific interests of both Northern and Southern cities in urban partnerships. The protagonists in the Pune-Bremen partnership have addressed this challenge by delegating decision-making powers and financial responsibility for some of Bremen's regular partnership funding to the partnership actors from Pune. This decision has reduced transaction costs for the partnership actors from Bremen and at the same time helped ensure that the partnership actors from Pune could set up projects that served their own interests. In the tramway project the municipal corporations of Pune and Pimpri Chinchwad even agreed to cover the costs for the Detailed Project Report; thereby preparing the ground for a business-oriented partnership project on a more equal footing. Although Pune eventually rejected to transfer the funds and the partnership initiative could not be implemented, the approach to sharing funding responsibilities for joint projects may be an important lesson in realising greater mutuality in transnational urban partnerships.

The recommendation to focus on analysing and addressing the vested interests of all cities involved in transnational urban partnerships is also strengthened by the increasingly dominant discourse on the need to implement climate policy by leveraging its economic, social and environmental co-benefits. In many cities, climate mitigation and adaptation have not yet become popular local policy concerns. While the relevance and potential threats of global climate change are increasingly recognised at a city level in both the Global North and South, the urban political agenda often remains dominated by topics that are perceived to be more urgent, such as economic growth, infrastructure development or energy security. A crucial strategy for boosting climate change's presence on the urban agenda is therefore to identify and publicise solutions that simultaneously address climate change and other, more dominant economic, social and environmental policy topics. The focus on such co-benefits is particularly important in Indian cities, given their limited financial and personnel resources, plus their prioritisation of economic growth and poverty reduction over climate change (Doll et al, 2013; Puppim de Oliveira, 2013; Sharma and Tomar, 2010). In India the perception that climate change is "relatively irrelevant to domestic politics" is widely shared (Dubash, 2012b, 5-6), as even without any mitigation efforts, in the medium-term India's per capita emissions will still remain below the global average (Fisher, 2012) This means that climate policy can only succeed when it is closely linked to practical co-benefits. This is exemplified in the fact that the co-benefits approach is a dominant theme in India's National Action Plan on Climate Change (NAPCC), "with mitigation understood to be the secondary benefit emerging from development policies." (IPCC, 2014b, 1152).

In German municipalities climate mitigation measures are also more likely to be introduced and implemented when their proponents can demonstrate positive side effects for environmental protection or local economic and social development and when they link climate policy to urban land-use planning and municipal energy management (Beermann, 2009; Deutsches Institut für Urbanistik, 2011).

Indian-German urban climate partnerships are therefore well advised to systematically analyse and incorporate co-benefits into the design of climate mitigation projects. In doing so partnerships do however face the additional challenge associated with the reality that co-benefit targets can differ considerably between actors from India and Germany. In European cities, strategies for sustainable and low carbon development often focus on environmental aspects whereas in Indian cities, economic and social development is often prioritised over environmental protection (Atteridge et al., 2009; Ghosh, 2014). Transnational urban partnership projects are far more likely to succeed if they address challenges in fields in which both partner cities can identify co-benefits that serve their respective interests. The major transformations to energy structures currently taking place in both Germany and India may therefore offer potential for future urban exchange and collaboration. Renewable energy deployment and energy efficiency measures offer immense opportunities to simultaneously realise economic and social development, environmental improvements and climate change mitigation.

7.2 Academic Implications

This study contributes to the development of theoretical and methodological approaches to transnational urban cooperation. Its research design is based on a combination of complementary theoretical concepts. From these concepts a set of explanatory variables have been deducted and operationalised in an index system. This approach has provided several theoretical and methodological insights and lessons.

The study helps provide a better understanding of the role of transnational climate governance networks in global attempts to address climate change, and it supplements efforts in other academic literature to set up typologies of this phenomenon. The work refers to the typology of transnational climate governance networks by Andonova et al. (2009) as a starting point from which to classify the distinct institutional forms of transnational urban cooperation as either public, hybrid public and private, or private transnational governance networks. It applies the typology's institutional classification of transnational climate governance networks as a criterion for case selection, to cover and compare the institutional diversity of transnational urban partnerships, based on their dominant actor structure. Accordingly, one primarily public cooperation (W2E), one hybrid public-private partnership (LRMCN) and two private-actor driven collaborations (DEWATS and tramway) were selected.

The findings of the comparative case study analysis suggest that it may be advisable to refine the typology's categorisation. In particular, the categories of "hybrid" and "private" require further specification. In fact, this study proposes to break down both categories. In Andonova et al.'s typology, the category of "hybrid" networks encompasses networks of mixed public *and* private actors as well as networks dominated by institutions which cannot be clearly classified according to the traditional categories of public *or* private, such as TMNs. These two forms can, however, differ considerably in their approaches towards setting up transnational collaborative projects. The TMN-facilitated LRMCN partnership between Nagpur and Freiburg, for example, represents a more top-down and externally-induced form of cooperation. Mixed public and private-actor driven partnerships, on the other hand, could be exclusively driven by local actors from the bottom up and without any external support. This study therefore proposes to replace the category of "hybrid" with the categories of "mixed public and private" networks and networks driven by "new transnational institutions".

Similarly, the case study's findings point to the need to refine the category of "private" transnational governance networks. In Andonova et al.'s typology "private" networks include both NGO and business-driven initiatives. A joint categorisation ignores the fact that NGO- and business-dominated transnational networks tend to be driven by different motives. This study's findings suggest that business-oriented partnerships may find it easier to identify and address the intrinsic interests of both the Southern and Northern partners and thereby facilitate greater mutuality in partnerships (as demonstrated by the Pune-Bremen tramway project). This is unsurprising in view of the assumption that a good business deal should be advantageous for both business partners. NGO-driven transnational cooperation often follows a different logic and is not necessarily driven by identifying intrinsic interests in both partner regions, as the DEWATS case in this study shows. Consequently, this study recommends replacing the category of "private" networks with the two categories of "NGO-driven" and "business-driven" transnational governance networks. These suggestions could benefit from further exploration in future research.

This study also builds upon Andonova et al.'s typology by exploring the enabling conditions for and barriers to transnational climate governance for which the typology does not offer any guidance. The study applies the theoretical concepts of policy transfer, institutionalism, policy entrepreneur, and social capital to explore the conditions for the success and failure of transnational climate governance at the urban level. The selection and integration of the theoretical perspectives in the research design has proven valuable. Each of the concepts has strong individual explanatory power for the analysis of transnational urban partnerships. Moreover, the theoretical concepts address each other's conceptual shortcomings and thereby offering a more comprehensive picture of the conditions for success and failure in urban climate partnerships.

The study refers to policy transfer theory to investigate the processes and conditions for knowledge exchange in transnational urban partnerships. It thereby addresses prevailing gaps in the policy transfer literature, such as on transfers at the subnational level and transfers across the distinct contexts of the Global North and South. The comparative analysis of the four case studies suggests that future research on policy transfers via transnational urban cooperation should focus on two areas in particular. The first area is the role of external moderators. The results of this study confirm Mossberger and Wolman's (2002) conclusion that moderators with in-depth insights into the original and borrowing settings can be crucial actors in facilitating policy transfers. This study finds that this conclusion is also applicable to the context of transnational urban cooperation. Especially when collaborative projects are set up in an ad hoc manner and not part of a long-term city partnership, knowledgeable external moderators can help translate between the different socio-political and cultural settings of the cities involved. In the two partnership projects in this study that were driven by external moderators, the moderators took on an additional, highly important role. ICLEI and the GIZ were instrumental in linking the projects to state institutions thereby helping them access state funds and gain necessary approval from state bodies. The results of this study also show that partnerships driven by external moderators may find it difficult to develop social capital between the actors involved and a dependence on external moderators may hamper the long-term development of urban partnerships.

This study identifies a focus on intrinsic interests as a second important condition for the transfer of know-how and policies in transnational urban cooperation. It concludes that if partnership projects address the vested interests of both partner cities involved, they are less likely to result in a one-sided flow of knowledge and funds. This focus on intrinsic interests speaks to an area that is still underdeveloped in the policy transfer literature; the role of the producers and senders of information in policy transfers (Wolman and Page, 2002). This study confirms that policy transfer literature needs to deepen its analysis of how to motivate senders to take part in policy transfers. With regard to urban climate cooperation involving cities from the Global North and South the study proposes to link policy transfer theory to debates on the co-benefits concept which is a useful approach to identifying and addressing the distinct vested interests of partner cities from different contexts.

As for existing approaches to categorising different forms and degrees of policy transfers (Rose, 1993; Dolowitz and Medearis, 2009; Dolowitz and Marsh, 2000) the study suggests that a direct transfer of policies between cities from the Global South and North appears to be unfeasible. The political framework conditions, social and economic challenges as well as the climatic conditions appear to be too distinct to allow for direct and complete policy transfers in the fields of climate protection and low carbon development. This is at least the case for German and Indian cities, the research units of this study. Transfers are more likely to be realised in the form of sharing and exchanging experiences in improving administrative processes, political agenda setting, and technological know-how. Dolowitz and Marsh (2000) categorise this form of transfer as "emulation", or the "transfer of the ideas behind the policy or program" (ibid., 13). Another important form that policy transfers between cities of the Global South and North can take is what Dolowitz and Marsh sum up as "inspiration for policy change" (ibid.). This refers to transferring the motivation to incorporate climate change and low carbon development into urban planning strategies something that is still a relatively novel approach in many cities, especially in countries of the Global South. The example of Nagpur in the LRMCN partnership demonstrates that transnational urban cooperation can serve as an important trigger for climate policy-making, with Nagpur becoming the first Indian city to introduce a city-wide clean energy policy in 2007 because of the LRMCN partnership project.

The study applies three additional theoretical approaches to ensure that the roles of structure (institutionalism) as well as individual and collective agency (policy entrepreneur and social capital theories) are addressed in the investigation of enabling conditions and barriers for transnational urban cooperation.

In its analysis of the effectiveness of different institutional forms of transnational urban cooperation this study refers to institutionalism literature. It tests whether a finding from a study on transnational partnerships between subnational states by Ralston (2013) who concludes such partnerships benefit from being formally institutionalised in the state legal system, is also applicable to transnational urban cooperation. Ralston finds that linking partnerships with state institutions helps ensure their long-term sustainability and reduces their dependency on partnership champions. The results of the analysis of the four German-Indian urban climate partnerships suggest that Ralston's findings do not necessarily apply to transnational urban cooperation. Rather this study shows that the long-term durability of transnational urban partnerships is facilitated by partnership entrepreneurs and a high degree of social capital among participants, rather than by connecting the partnerships with state institutions. The four cases analysed in fact point to a negative correlation between the formal involvement of state institutions and the partnerships' long-term durability. The study does, however, conclude that for the implementation of larger-scale partnership projects, close linkages with state institutions may be necessary to secure the state approval required. Future research should focus on whether the challenges identified in securing state approval in India, such as bureaucratic hurdles, delayed implementation and loss of partnership social capital, are also evident in partnerships featuring cities from other countries, for example those with less centralised and hierarchical political systems.

Deriving variables from the policy entrepreneur and social capital literature the study investigated the role of individual and collective agency in transnational urban cooperation. The case studies analysed suggest that policy entrepreneurs driving urban North-South cooperation, termed in this study 'partnerships entrepreneurs', are particularly important in the founding and maintaining of partnerships. Partnerships driven by partnerships entrepreneurs are less dependent on local political decision makers and their limited terms of office. At the same time the study confirms the finding by Mintrom (1997) that non-state entrepreneurs can still find it difficult to gain legislative approval from political decision makers for their projects, even if the entrepreneurs have good access to political circles. This study suggests that external moderators may be more influential in achieving political approvals as compared to partnership entrepreneurs.

As partnership entrepreneurs only played a major role in two of the four partnership projects analysed, more research is needed to identify wider trends in the characteristics and capacities required of policy entrepreneurs in transnational urban partnerships. This research would also deepen international and transnational perspectives on the concept of policy entrepreneurs. Research on the role of policy entrepreneurs in international or transnational projects is still underdeveloped (an exception being Ralston (2013)). Broadening this field of research is of importance due to the increasingly global scope of policy issues, such as climate change.

The same conclusion can be drawn with regard to the social capital concept. Like the policy entrepreneur concept, theory development on social capital has been mainly based on an analysis of individual communities, viewed in isolation. Since the beginning of the 21st century cities have become increasingly integrated in international and transnational networks (Campbell, 2012). As Krishna and Shrader (2000) rightly point out, it is therefore instructive to explore the role of social capital in such networks and discuss its potential implications for social capital theory development. This study adds to the still limited but emerging research on the role of social capital in trans-urban learning and cooperation. It supports Bontenbal and van Lindert's (2008) and Campbell's (2012) finding that social capital has a crucial function in the stabilisation and long-term maintenance of city partnerships. The Pune-Bremen cooperation analysed in this study demonstrates that a high degree of social capital can help maintain partnership relationships and projects during local political leadership transitions, periods which often pose a threat to the continuity of urban partnerships. In urban South-North partnerships social capital can also reduce the dependency on the intercultural capacities of external moderators to translate between the cities' distinct contexts. Putnam et al.'s (2003) distinction between "bridging" and "bonding" forms of social capital may help broaden the concept of social capital within the international and transnational context. This study concludes that in urban North-South partnerships, "bridging" social capital in particular is required to bring city actors from different cultural and socio-economic contexts together. According to Putnam et al. "bridging" social capital evolves in heterogeneous networks whose members have diverse socio-economic backgrounds. They argue that "bridging" social capital is difficult to develop, but it is vital for the development of resilient and inclusive communities and for facilitating social cooperation in increasingly complex societal settings. This study suggests that when applying the term to transnational urban cooperation the definition of "bridging" social capital should be extended to

7.2 Academic Implications

include factors such as equality and intercultural competence. Both factors are prerequisites for developing and maintaining trustful and productive partnership relations between city actors. Equality within relationships and an understanding of the distinct perceptions, discourses and interests of the respective partner cities' actors are particularly important in collaborative urban projects addressing climate change, as the topic remains highly contested and prone to misunderstandings, especially between actors from the Global South and North.

In the study the explanatory variables derived from the theoretical concepts above were operationalised in an index system. This index system offers a methodological tool with which to analyse the enabling conditions and barriers associated with transnational urban cooperation. The system thereby facilitates the identification and assessment of correlations between explanatory variables and the dependent variable, i.e. transnational urban partnership success, in the comparative case study analysis. The index system also helps identify correlations between individual explanatory variables. These correlations then pave the way for the exploration of linkages between the theoretical concepts from which the explanatory variables were derived. Applied to Indian-German urban climate cooperation, the index system points to several correlations between explanatory variables. For example, it reveals a potential negative correlation between the integration of partnerships into state institutions and the development of social capital in such partnerships. Recognising that the interplay between political institutions and social capital is generally under-researched (Heydenreich-Burck, 2010), this finding offers a new perspective and a potential starting point for linking the theoretical concepts of institutionalism and social capital. Similarly, the negative correlation between the degree to which partnerships are institutionalised in the state system and the involvement of partnership entrepreneurs supports demands by Kingdon (2011) to incorporate both structure (institutionalism) and agency (policy entrepreneur concept) as explanatory variables of social phenomena and analyse their interlinkages, rather than treating them as exclusive explanatory approaches. The positive correlation between the indexes of partnership entrepreneurs and social capital confirms existing research highlighting the key role that policy entrepreneurs can play in the development of social capital (Putnam et al., 2003).

The index system refers to Tucker's (2010) "grounded" indicator approach which aims to foster the development of rigorous, but flexible measurement tools for complex social phenomena in different settings and times, by incorporating and integrating theoretical and empirical insights. To explore the index system's theoretical validity and its wider applicability future research is invited to test, if and how the index system needs to be adjusted to analyse urban South-North partnerships between countries other than Germany and India. It would be beneficial to test whether the index system can also be applied to analyse the conditions for success and failure in urban South-South or North-North partnerships. Moreover, the index system's application to other cases could serve as a basis for a discussion about the weight individual indicators should be assigned within the indexes. This was not covered in this study due to a lack of suggestions on index weighting in the literature and empirical data from which the indicators were derived.

Based on this study's key findings it would be advisable for future research to investigate whether the assumed trade-off between bottom-up and top-down development is also found in other transnational urban partnership initiatives and if methods exist to overcome this conflict. It would be insightful to identify urban partnerships that combine small-scale civil society initiatives with large-scale, formally institutionalised public projects and explore if these partnerships are able to both realise successful large-scale projects and maintain long-term collaboration. More research is also required on how to best integrate city-level climate action and urban cooperation into national climate strategies, while at the same time preserving cities' important role as laboratories for experimentation in the search for innovative and feasible solutions to climate change. Another task for future research is to take a more in-depth look at the role and relevance of policy windows in transnational urban cooperation. This study's index system does not offer any clear results with regard to the importance of policy windows in the four cases analysed, policy windows were therefore not selected as a focus area in the comparative analysis. The individual case studies do suggest that policy windows may have played a substantial role in the processes and outcomes of all four projects. In fact, the problem and policy streams were relatively advantageous and stable throughout the course of the projects, whereas the political stream was more dynamic (at times advantageous and at times disadvantageous). The political stream thus appears to have been a crucial factor, as during the advantageous periods of the stream (e.g. the partnership MoUs and the Rio Earth Summit in the DEWATS project; initial approval from the local commissioner in the tramway project; national funding schemes and support from the local commissioner in the W2E project; and the emerging interest of the Government of India in developing urban clean energy strategies in the LRMCN), the partnership projects were usually making progress. During adverse periods of the political stream policy windows closed, leading to deadlocks or even the failure of these projects. Based on these observations, future studies may wish to consider a refinement of indicator c(1) assessing the role of policy windows in the index system. Finally, several critical reflections and lessons on persisting difficulties in addressing the above outlined Northern bias in urban studies can be drawn (see section 4.1). This study confirms suggestions by post-colonial urban researchers that investigating urban city con-

7.2 Academic Implications

nections and relations is an insightful approach towards strengthening cosmopolitan perspectives in urban studies. Studying urban South-North relations inherently demands that the researcher reflects upon the similarities and differences between the perspectives of Northern and Southern cities to integrate them into the design of their research. It is highly recommendable to conduct research visits and interviews in both the Southern and Northern cities involved. Northern researchers can particularly benefit from longer term and if possible repeated stays in their Southern case study cities and Southern researchers in Northern cities to gain first-hand insight into the respective contextual conditions there.

Selecting an adequate theoretical and methodological design can greatly facilitate research that aims to bridge and integrate the perspectives of both Southern and Northern cities. This study applies Tucker's (2010) suggestion of strengthening "grounded" indicator approaches as a promising method of studying urban South-North collaboration. The "grounded" index system offers the flexibility to combine initial empirical insights on situational contextual conditions with more general theoretical assumptions, which enables the researcher to ensure the inclusion of both Northern and Southern perspectives and sources.

The selection of a suitable research topic and theoretical-methodological framework does however not ensure that the researcher remains wholly reflective and self-critical with regard to persisting North-South biases. Unfortunately, these are manifold and sometimes subtle in form.

In this study of four cases of Indian-German urban collaboration, potential biases in the data cannot be fully ruled out, specifically regarding the possibility that interview partners from Indian cities may not want to be too critical of their partners from Germany. This avoidance of criticism may stem from the fact that some Southern partners feel that they need to maintain good relations with their Northern counterparts in view of potential future funding.

Another potential bias can arise from a lack of equality in the interview partners and interview data selected for analysis. An example of this may be found in the DEWATS case in this study which contains more references to interviews with German partners than with Indian partners. An explanation for this unbalanced representation of German and Indian interview partners may be because the two major Indian protagonists directly involved in the preparation and implementation of the DEWATS project were not available for interview, as they had passed away a few years before the study was conducted. Another explanation may be the possibility that interviewees from the North found it easier to understand the interview questions and therefore gave responses that better met the author's research needs. This leads to the conclusion that cultural reception should be considered when conducting research with interviewees from a different cultural background to the interviewer. An interesting area of future research would be to investigate the extent to which an interviewee's responses and an interviewer's reception and perception of responses may be grounded in their cultural background and to explore methods that ensure these cultural differences do not lead to bias in qualitative research.

A third example in this study which shows the need for extra reflection in terms of North-South biases in the collection and use of data, is the definition of target groups in transnational urban cooperation. In this study, the researcher defined the target groups based on interview responses and documentation suggesting that in all four projects the target groups were solely located in the Indian cities, usually formed of (sections of) the local population or the city administration in these cities. With regard to overcoming North-South divides in urban research and pursuing greater mutuality in South-North urban partnerships more generally it may be insightful to challenge these assumptions and place greater emphasis on the identification of potentially overlooked target groups in German cities and their sometimes less obvious and tangible benefits, such as individual and organisational learning; the development of solidarity and social capital; or the empowerment of NGOs, public administrations, or businesses.

These examples demonstrate the immense challenges associated with overcoming North-South biases in research and practise. This should, however, not discourage researchers from continuously aspiring to achieve the ideal of a truly cosmopolitan form of urbanism. Rather such efforts and the exploration of how bias occurs and can be avoided offers great potential for better overcoming such bias in future research.

8 References

Acuto, M. (2013). City Leadership in Global Governance. *Global Governance: A Review of Multilateralism and International Organizations*, 19(3), 481–498.
Adelphi Consult. (2012). *Report for Technical Learning Visit "Support to National Urban Sanitation Policy" September 23th to October 3rd 2012 Hamburg and Berlin, Germany.*
Agarwal, A., & Narain, S. (1991). *Global Warming in an Unequal World. A case of Environmental Colonialism.* New Delhi.
Alber, G., & Kern, K. (2008). Governing Climate Change in Cities: Modes of Urban Climate Governance in Multi-level Systems. *Competitive Cities and Climate Change, OECD Conference Proceedings, Milan, Italy, 9-10 October 2008*, 171–196.
Andonova, L. B., Betsill, M. M., & Bulkeley, H. (2009). Transnational climate governance. *Global environmental politics*, Vol. 9(2), 52–73.
Anguelovski, I., & Carmin, J. (2011). Something borrowed, everything new: innovation and institutionalization in urban climate governance. *Current Opinion in Environmental Sustainability*, 3(3), 169–175. doi:10.1016/j.cosust.2010.12.017
Ansell, C., & Gash, A. (2007). Collaborative Governance in Theory and Practice. *Journal of Public Administration Research and Theory*, 18(4), 543–571. doi:10.1093/jopart/mum032
Atteridge, A., Nilsson Axberg, G., Goel, N., Kumar, A., Lazarus, M., Ostwald, M., … (2009). *Reducing greenhouse gas emissions in India: financial mechanisms and opportunities for EU-India cooperation*. Stockholm. Retrieved from http://www.sei-international.org/mediamanager/documents/Publications/Climate-mitigation-adaptation/reducinggreenhousegasemissions-india.pdf (19-02-2016)
Augustin, K., Giese, T., & Dube, R. (2010). *Waste to Energy for Urban India through Co-fermentation of Organic Waste and Septage: ISWA World Congress*. Hamburg. Retrieved from http://www.iswa.org/uploads/tx_iswaknowledgebase/Augustin.pdf (19-02-2016)
Aylett, A. (2010). Conflict, Collaboration and Climate Change: Participatory Democracy and Urban Environmental Struggles in Durban, South Africa. *International Journal of Urban and Regional Research*, 34(3), 478–495. doi:10.1111/j.1468-2427.2010.00964.x
Bäckstrand, K. (2008). Accountability of Networked Climate Governance. The Rise of Transnational Climate Partnerships. *Global environmental politics*, 8(3), 74–102.
Beermann, J. (2009). *100% Renewable Energy Regions in Europe. A comparative analysis of local renewable energy development: Master's Thesis.* Freie Universität Berlin.
Beermann, J. (2014). Urban partnerships in low-carbon development: Opportunities and challenges of an emerging trend in global climate politics. *URBE - Revista Brasileira de Gestão Urbana*, 6(541), 170–183. doi:10.7213/urbe.06.002.SE03
Beermann, J. & Tews, K. (2015). Preserving Decentralised Laboratories for Experimentation under Adverse Framework Conditions - Why Local Initiatives as a Driving Force for Germany's Renewable Energy Expansion Must Reinvent Themselves. *FFU Report 03-2015.* Retrieved from http://www.polsoz.fu-berlin.de/polwiss/forschung/systeme/ffu/aktuell/15_juli-ffu-report.html (19-02-2016)
Berkes, F. (2002). Cross-Scale Institutional Linkages: Perspectives from the Bottom Up. In E. Ostrom (Ed.), *The drama of the commons* (p 293–321). Washington, DC: National Academy Press.

Betsill, M. M., & Bulkeley, H. (2004). Transnational Networks and Global Environmental Governance: The Cities for Climate Protection Program. *International Studies Quarterly*, *48*(2), 471–493. doi:10.1111/j.0020-8833.2004.00310.x

Betsill, M. M., & Bulkeley, H. (2006). Cities and the Multilevel Governance of Global Climate Change. *Global Governance*, *12*(2), 141–159.

Betsill, M., & Bulkeley, H. (2007). Looking Back and Thinking Ahead: A Decade of Cities and Climate Change Research. *Local Environment*, *12*(5), 447–456. doi:10.1080/13549830701659683

Bhagat, R. B. (2005). Rural-Urban Classification and Municipal Governance in India. *Singapore Journal of Tropical Geography*, *26*(1), 61–73. doi:10.1111/j.0129-7619.2005.00204.x

Biermann, F., Abbott, K., Andresen, S., Backstrand, K., Bernstein, S., Betsill, M. M., Biermann, F., Abbott, K., Andresen, S., Backstrand, K., Bernstein, S., Betsill, M. M., … (2012). Navigating the Anthropocene: Improving Earth System Governance. *Science*, *335*(6074), 1306–1307. doi:10.1126/science.1217255

Bondre, A. (2005). *The Pune-Bremen Co-operation. An Inspiring Model of Sustainable Development*.

Bontenbal, M. (2009). Strengthening urban governance in the South through city-to-city cooperation: Towards an analytical framework. *Habitat International*, *33*(2), 181–189.

Bontenbal, M., & van Lindert, P. (2008). Bridging local institutions and civil society in Latin America: can city-to-city cooperation make a difference? *Environment and Urbanization*, *20*(2), 465–481. doi:10.1177/0956247808096123

Bontenbal, M., & van Lindert, P. (2009). Transnational city-to-city cooperation: Issues arising from theory and practice. *Habitat International*, *33*(2), 131–133. doi:10.1016/j.habitatint.2008.10.009

Bremen Overseas Research and Development Association - BORDA. (2010). *DEWATS – Decentralized Wastewater Treatment Solutions. Demand-based technical solutions to reduce waterpollution by small and medium enterprises and settlements in densely populated areas*. Retrieved from http://www.borda-net.org/dewats-service-packages/dewats-the-system.html

Bulkeley, H. (2006). Urban sustainability: learning from best practice? *Environment and Planning A*, *38*(6), 1029–1044. doi:10.1068/a37300

Bulkeley, H., & Betsill, M. M. (2013). Revisiting the urban politics of climate change. *Environmental Politics*, *22*(1), 136–154. doi:10.1080/09644016.2013.755797

Bulkeley, H., & Kern, K. (2006). Local Government and the Governing of Climate Change in Germany and the UK. *Urban Studies*, *43*(12), 2237–2259. doi:10.1080/00420980600936491

Bulkeley, H., & Newell, P. (2010). *Governing climate change*. London ;, New York: Routledge.

Bulkeley, H., Schroeder, H., Janda, K., Zhao, J., Armstrong, A., Chu, S. Y., & Ghosh, S. *Cities and Climate Change: The role of institutions, governance and urban planning: Report prepared for the World Bank Urban Symposium on Climate Change*. Retrieved from http://www.eci.ox.ac.uk/publications/downloads/bulkeley-schroeder-janda09.pdf

C40, UCLG, ICLEI (2014). *The Compact of Mayors. Goals, Objectives and Commitments*. Retrieved from http://www.iclei.org/fileadmin/user_upload/ICLEI_WS/Documents/advocacy/Climate_Summit_2014/Compact_of_Mayors_Doc.pdf (19-02-2016)

Campbell, T. (2012). *Beyond smart cities: How cities network, learn and innovate*. Abingdon, Oxon, New York, NY: Earthscan.

Castán Broto, V., & Bulkeley, H. (2013). A survey of urban climate change experiments in 100 cities. *Global Environmental Change*, *23*(1), 92–102. doi:10.1016/j.gloenvcha.2012.07.005

Chakrabarty, D. (2002). *Habitations of modernity: Essays in the wake of subaltern studies*. Chicago: University of Chicago Press.

Chandra, L. *Nagpur Local Renewable Activities and City Energy Status Report. Presentation at Local Renewables Model Communities Network Project International Workshop in Nagpur, November 28-30, 2006*.

8 References

Climate Alliance - Klima-Bündnis. (2008). *Solutions for Change: How Local Governments are Making a Difference in Climate Protection*.
Climate Alliance - Klima-Bündnis. (2012). *European Funds for local Climate Protection*. Retrieved from http://www.klimabuendnis.org/fileadmin/inhalte/dokumente/2012/1207_EU-Funding_website.pdf (19-02-2016)
Coleman, J. S. (1990). Foundations of social theory. Cambridge, Mass: Belknap Press of Harvard University Press.
Consult Team Bremen, G. I. I. (2007). *Detailed Project Report on Tramways*. Bremen.
Corfee-Morlot, J., Kamal-Chaoui, L., Donovan, M. G., Cochran, I., Robert, A., & Teasdale, J. (2009). *Cities, Climate Change and Multilevel Governance: OECD Environmental Working Papers N° 14, 2009*.
Dawes, R. M. (1980). Social Dilemmas. *Annual Review of Psychology*, 31(1), 169–193.
Denters, B., & Mossberger K. (2006). Building Blocks for a Methodology for Comparative Urban Political Research. Urban Affairs Review, 41(4), 550–571. doi:10.1177/1078087405282607
Deutsche Gesellschaft für Internationale Zusammenarbeit (GIZ). *Write up on Waste to Energy Project for Standing Committee*.
Deutsche Gesellschaft für Technische Zusammenarbeit (GTZ). (2010). *Workshop on Waste Management at Nashik Local Center, Nashik October 19, 2010. Workshop Report*.
Deutscher Städtetag. (2010). *Positionspapier Klimaschutz in Städten*.
Deutsches Institut für Urbanistik. (2011). *Klimaschutz in Kommunen - Praxisleitfaden*. Retrieved from http://www.leitfaden.kommunaler-klimaschutz.de/download.html
Devers-Kanoglu, U. (2009). Municipal partnerships and learning – Investigating a largely unexplored relationshi *City-to-City Co-operation*, 33(2), 202–209. doi:10.1016/j.habitatint.2008.10.019
Dhakal, S. (2004). *Urban energy use and greenhouse gas emissions in Asian mega-cities: Policies for a sustainable future*. Kitakyushu, Japan: Urban Environmental Management Project, Institute for Global Environmental Strategies.
Dhakal, S. (2006). *Urban Transportation and the Environment in Kathmandu Valley, Nepal: Integrating Global Carbon Concerns into Local Air Pollution Management*. Retrieved from http://www.energycommunity.org/documents/iges_start_final_reprot.pdf (19-02-2016)
Dhakal, S. (2009). Urban energy use and carbon emissions from cities in China and policy implications. *Energy Policy*, 37(11), 4208–4219. doi:10.1016/j.enpol.2009.05.020
Dodman, D. (2009). Blaming cities for climate change? An analysis of urban greenhouse gas emissions inventories. *Environment and Urbanization*, 21(1), 185–201. doi:10.1177/0956247809103016
Dolowitz, D., & Marsh, D. (2000). Learning from Abroad: The Role of Policy Transfer in Contemporary Policy-Making. *Governance*, 13(1), 5–23. doi:10.1111/0952-1895.00121
Dolowitz, D., & Medearis, D. (2009). Considerations of the obstacles and opportunities to formalizing cross-national policy transfer to the United States: a case study of the transfer of urban environmental and planning policies from Germany. *Environment and Planning C: Government and Policy*, 27(4), 684–697. doi:10.1068/c0865j
Dubash, N. K. (Ed.). (2012a). *A handbook of climate change and India: Development, politics, and governance*. Abingdon, Oxon ;, N.Y: Earthscan.
Dubash, N. K. (2012b). Climate Politics in India. Three narratives. In N. K. Dubash (Ed.), *A handbook of climate change and India. Development, politics, and governance* (p 197–205). Abingdon, Oxon ;, N.Y: Earthscan.
Dubash, N. K. (2012c). *The Politics of Climate Change in India: Narratives of Equity and Co-benefits: Centre for Policy Research Climate Initiative. Working Paper 2012/1*.
Dube, R. (2012). *The International Climate Initiative (ICI). Producing Energy from Waste and Sewerage. Brochure*.
Dube, R. (2013, January). *Waste to Energy. Co-Fermentation of Kitchen Waste and Septage: Presentation*.

Dudwick, N., Kuehnast, K., Nyhan Jones, V., & Woolcock, M. (2006). *Analyzing social capital in context: A guide to using qualitative methods and data. WBI working papers.* Washington, D.C: World Bank Institute.

ENGAGEMENT GLOBAL – Service for Development Initiatives, Service Agency Communities in One World (2014). *50 Municipal Climate Partnerships by 2015. Documentation of the second phase of the project. Dialog Global 32.* Retrieved from http://www.service-eine-welt.de/en/climatepartnerships/climatepartnerships-start.html (19-02-2016)

Fisher, S. (2012). Policy storylines in Indian climate politics: opening new political spaces? *Environment and Planning C: Government and Policy, 30*(1), 109–127. doi:10.1068/c10186

Forum Städtesolidarität Breme-Pune e.V. (2006). *30 Jahre Förderung deutsch-indischer Freundschaft und Zusammenarbeit.*

Freie Hansestadt Bremen. (2005). *Partner in vielen Bereichen: Indien und Bremen. Zusammenarbeit für nachhaltige Entwicklung: Aus der Praxis der Entwicklungszusammenarbeit Bremens.* Retrieved from http://www.ez.bremen.de/sixcms/media.php/13/indien-bremen_n_korrektur.pdf (19-02-2016)

Gaikwad, S. (2013, October 10). Overflowing sewage floods housing societies in Chikhli. *The Times of India, Pune.* Retrieved from http://timesofindia.indiatimes.com/city/pune/Overflowing-sewage-floods-housing-societies-in-Chikhli/articleshow/23859121.cms (19-02-2016)

Gale, N., Heath, G., Cameron, E., Rashid, S., & Redwood, S. (2013). Using the framework method for the analysis of qualitative data in multi-disciplinary health research. *BMC Medical Research Methodology, 13*(1), 117.

George, A. L., & Bennett, A. (2005). *Case studies and theory development in the social sciences.* Cambridge, Mass. [u.a.]: MIT Press.

Ghosh, B. (2014). Sustainability appraisal of emerging trajectories in solar photovoltaic and urban mobility systems in India and Thailand: Master's thesis. Eindhoven University of Technology. Retrieved from http://repository.tue.nl/780502

Government of India. Ministry of Urban Development. (2015). *Smart Cities. Mission Statement & Guidelines.* Retrieved from http://smartcities.gov.in/writereaddata/SmartCityGuidelines.pdf (19-02-2016)

Gujar, A. S. (2010). *Maharashtra Arogya Mandal. Decentralized Wastewater Treatment System (DEWATS).* Pune.

Hafteck, P. (2003). An introduction to decentralized cooperation: definitions, origins and conceptual mapping. *Public Administration and Development, 23*, 333–345.

Hakelberg, L. (2011). *Governing climate change by diffusion: Transnational municipal networks as catalysts of policy spread.* Berlin: Freie Univ. Berlin Forschungszentrum für Umweltpolitik.

Hamburg Wasser Kompetenznetzwerk. (2010). *Pilot Project on "Waste to Energy Project in Nashik - Energy production from sewage and organic waste in Nashik". Feasibility Study.* Hamburg.

Hanifan, L. J. (1916). The rural school community center. *Annals of the American Academy of Political and Social Science 67*, 130–138.

Hardoy, J., & Romero Lankao, P. (2011). Latin American cities and climate change: challenges and options to mitigation and adaptation responses. *Current Opinion in Environmental Sustainability, 3*(3), 158–163. doi:10.1016/j.cosust.2011.01.004

Heinz, W. & Leitermann, W. (2004). Kooperationsbeziehungen zwischen deutschen Städten und Kommunen in Entwicklungsländern. In Bundeszentrale für politische Bildung. (Ed.), *Aus Politik und Zeitgeschichte: Bundeszentrale für politische Bildung, B 12–16/2004, Beilage zur Wochenzeitung "Das Parlament".* Bonn.

Hertle, H. & Schächtele, K. (2008). *Who's ahead? Climate Cities Benchmark in Japan, U.S. and Germany.* Paper presented at ACEEE Conference, August 2008.

8 References

Heydenreich-Burck, K. (2010). *Politische Institutionen und Sozialkapital: Ein internationaler Vergleich der Determinanten sozialen Kapitals*. Frankfurt, M, Berlin, Bern, Bruxelles, New York, NY, Oxford, Wien: Lang.
Hilliges, G. (2006). 30 Years of Partnership Between Bremen and Pune. In K. Voll & D. Beierlein (Eds.), *Berliner Studien zur internationalen Politik und Gesellschaft: Vol. 3. Rising India - Europe's partner? Foreign and security policy, politics, economics, human rights and social issues, media, civil society and intercultural dimensions* (p 1105–1114). New Delhi, Berlin: Mosaic Books; Weissensee Verlag.
Hodson, M., & Marvin, S. (2010). Urbanism in the anthropocene: Ecological urbanism or premium ecological enclaves? *City: Analysis of Urban Trends, Culture, Theory, Policy, Action, 14*(3), 298–313. doi:10.1080/13604813.2010.482277
Holzinger, K., Jörgens, H., & Knill, C. (Eds.). (2007). *Transfer, Diffusion und Konvergenz: Konzepte und Kausalmechanismen*. Wiesbaden: VS Verlag für Sozialwissenschaften.
Hooghe, L., & Marks, G. (2001). Types of Multi-Level Governance. *European Integration online Papers (EIoP), Vol. 5*(No. 11). Retrieved from Available at SSRN: http://ssrn.com/abstract=302786 or http://dx.doi.org/10.2139/ssrn.302786
ICLEI. (2007). *Local Renewables Model Communities Network. Year in Review 2007*. Retrieved from http://local-renewables.iclei.org/fileadmin/template/projects/localrenewables/files/Local_Renewables/Publications/LocalRenewables_Review2007.PDF (19-02-2016)
ICLEI. (2008). *Addendum to Local Renewables Year in Review 2007*. Retrieved from http://local-renewables.iclei.org/fileadmin/template/projects/localrenewables/files/Local_Renewables/Publications/LR_addendum_v2_26May08_smal.pdf (19-02-2016)
ICLEI. (2009). *Freiburg im Breisgau, Germany. Long-term strategies for climate protection in Green City Freiburg: Case Study 104*. Retrieved from http://www.iclei.org/fileadmin/PUBLICATIONS/Case_Studies/ICLEI_cs_104_Freiburg_June_2010.pdf (19-02-2016)
ICLEI. (2010a). *City Completion Report: Nagpur, India. Local Renewables Model Community Network*. Delhi.
ICLEI. (2010b). *Local Renewables project, Indian Country Report*. Delhi.
ICLEI. (2010c). *Nagpur, India. Water sector audit enables efficient use of water and energy resources in Nagpur: Case Study 110*. Retrieved from http://www.indiaenvironmentportal.org.in/content/322814/water-sector-audit-enables-efficient-use-of-water-and-energy-re-sources-in-nagpur/
ICLEI. (2013). *ICLEI Mayors discuss with Ban Ki-Moon and ministers multi-level climate actions and support at Cities Day of COP19 in Warsaw*. Retrieved from http://hosted.verticalre-sponse.com/413987/0a170817c4/1626003033/627d49209f/ (19-02-2016)
International Office Agenda 21. *Shaping the Future*. Pune.
IPCC. (2014a). Climate Change 2014. Synthesis Report Summary for Policymakers. Retrieved from http://www.ipcc.ch/ (19-02-2016)
IPCC. (2014b). *Climate Change 2014: Mitigation of Climate Change. Contribution of Working Group III to the Fifth Assessment Report of the Intergovernmental Panel on Climate Change [Edenhofer, O., R. Pichs-Madruga, Y. Sokona, E. Farahani, S. Kadner, K. Seyboth, A. Adler, I. Baum, S. Brunner, Eickemeier, B. Kriemann, J. Savolainen, S. Schlömer, C. von Stechow, T. Zwickel & Minx, J.C. (eds.)]*. Cambridge, New York. Retrieved from https://www.ipcc.ch/report/ar5/wg3/ (19-02-2016)
Isailovic, M., Widerberg, O., & Pattberg, P. (2013). Fragmentation of global environmental governance architectures: a literature review. *Available at SSRN 2479930*,
Jadhav, R. (2013). Study finds cities like Pune grappling with air pollution: Private vehicles rise as public transport remains poor. *The Times of India*, October 7, Pune, 4.
Jänicke, M. (2013). *Accelerators of Global Energy Transformation: Horizontal and Vertical Reinforcement in Multi-Level Climate Governance: IASS Working Paper*. Potsdam.

Johnson, H., & Wilson, G. (2009). Learning and mutuality in municipal partnerships and beyond: A focus on northern partners. *City-to-City Co-operation, 33*(2), 210–217. doi:10.1016/j.habitatint.2008.10.013

Kabbert, R. (2014, March 20). Highlight in langer Partnerschaft. Konzerte des Sinfonieorchesters der Musikschule Bremen im indischen Pune als gesellschaftliches Ereignis. *Weser Kurier.* Retrieved from http://www.weser-kurier.de/bremen/stadtteile/ausgabe-nordost_artikel,-Highlight-in-langer-Partnerschaft-_arid,808044.html (19-02-2016)

Kamal-Chaoui, L., & Robert, A. (2009). *Competitive Cities and Climate Change.* Paris: OECD Publishing.

Kern, K., & Bulkeley, H. (2009). Cities, Europeanization and Multi-level Governance: Governing Climate Change Through Transnational Municipal Networks. *Journal of Common Market Studies,*

King, G., Keohane, R. O., & Verba, S. (1994). *Designing social inquiry: Scientific inference in qualitative research.* Princeton, N.J: Princeton University Press.

Kingdon, J. W. (2011). *Agendas, alternatives, and public policies* (2nd ed.). *Longman classics in political science.* Boston, Mass. [u.a.]: Pearson, Longman.

Kithiia, J. (2011). Climate change risk responses in East African cities: need, barriers and opportunities. *Current Opinion in Environmental Sustainability, 3*(3), 176–180. doi:10.1016/j.cosust.2010.12.002

Klatte, G. (2005). *AWO – BORDA. Project Number: 04 / 419. Project Title: Project Implementation of DEWATS at AWO-Project-Partners, Maharashtra Arogya Mandal (MAM) and Matru Mandir (MM), in Maharashtra, India. Report of "Short-term-visit" to Pune and Devrukh / India. 9th of April to 17th of April 2005.*

Krause, R. M. (2011). Policy Innovation, Intergovernmental Relations, and the Adoption of Climate Protection Initiatives by U.S. Cities. Journal of Urban Affairs, 33(1), 45–60. doi:10.1111/j.1467-9906.2010.00510.x

Krishna, A., & Shrader, E. (2000). *Cross Cultural Measures of Social Capital: A Tool and Results from India and Panama: Social Capital Initiative Working Paper 21.* Washington, D.C.

Loury, G. C. (1977). A Dynamic Theory of Racial Income Differences. In P. A. Wallace & A. M. LaMond (Eds.), *Women, minorities, and employment discrimination* (pp. 153–188). Lexington, Mass: Lexington Books.

Loury, G. C. (1987). Why should we care about group inequality? *Social Philosophy and Policy, 5,* 249–271.

Marquardt, J., Steinbacher, K., Schreurs, M.A. (2015). Driving force or forced transition?, *Journal of Cleaner Production.* In Press. 1-12. doi: 10.1016/j.jclepro.2015.06.080

McFarlane, C. (2010). The Comparative City: Knowledge, Learning, Urbanism. *International Journal of Urban and Regional Research, 34*(4), 725–742. doi:10.1111/j.1468-2427.2010.00917.x

McKinsey Global Institute. (2010). *India's urban awakening: Building inclusive cities, sustainable economic growth.*

Medearis, D., & Dolowitz, D. (2013). Cross-National Urban Sustainability Learning: A Case Study on "Continuous Interaction" in Green Infrastructure Policies. In H. A. Mieg & K. Töpfer (Eds.), *Routledge studies in sustainable development: Vol. 1. Institutional and social innovation for sustainable urban development* (p 233–245). Abingdon, Oxon: Routledge.

Memorandum of Understanding between GTZ-ASEM and Nashik Municipal Corporation (2010).

Meyer-Timpe, U. (2010, March 3). Leere Kassen: Pleite und gelähmt. *ZEIT Online.* Retrieved from http://www.zeit.de/2010/10/Kommunalfinanzen.

Mintrom, M. (1997). Policy Entrepreneurs and the Diffusion of Innovations. *American Journal of Political Science, 41*(03), 738–770. doi:10.2307/2648004

Mintrom, M., & Vergari, S. (1998). Policy Networks and Innovation Diffusion: The Case of State Education Reforms. *The Journal of Politics, 60*(01), 126–148. doi:10.2307/2648004

Mossberger, K., & Wolman, H. (2003). Policy Transfer as a Form of Prospective Policy Evaluation: Challenges and Recommendations. *Public Administration Review*, *63*(4), 428–440. doi:10.1111/1540-6210.00306

Mukhopadhyay, P., & Revi, A. (2012). Climate change and urbanization in India. In N. K. Dubash (Ed.), *A handbook of climate change and India. Development, politics, and governance* (p 303–316). Abingdon, Oxon;, N.Y: Earthscan.

Mutz, D. (2012). *Indo-German Environment Partnership (IGEP) Programme.*

Nair, K. S. (2009). *An Assessment of the Impact of Climate Change on the Megacities of India and of the Current Policies and Strategies to meet Associated Challenges: Fifth Urban Research Symposium.*

Nakamura, H. (2010). *Political factors facilitating practice adoption through Asian intercity network programmes for the environment: IGES Discussion Paper.*

Nashik Municipal Corporation. (2011). *Draft City Sanitation Plan. Volume I – Main report. September 2011.*

Nijman, J. (2007). Introduction—Comparative Urbanism. *Urban Geography*, *28*(1), 1–6. doi:10.2747/0272-3638.28.1.1

Nitschke, U., Held, U., & Wilhelmy, S. (2009). Challenges of German city2city cooperation and the way forward to a quality debate. *Habitat International*, *33*(2), 134–140. doi:10.1016/j.habitatint.2008.10.023

Nyhan Jones, V., & Woolcock, M. (2010). Measuring the Dimensions of Social Capital in Developing Countries. In G. Walford, E. Tucker, & M. Viswanathan (Eds.), *The SAGE Handbook of measurement* (p 537–559). [S.l.]: Sage Publications.

OECD DAC Network on Development Evaluation. (2010). *Evaluating Development Co-operation. Summary of Key Norms and Standards*. Retrieved from http://www.oecd.org/dac/evaluation/summaryofkeynormsandstandards.htm (19-02-2016)

Ohlhorst, D., Tews, K., & Schreurs, M. A. (2013). Energiewende als Herausforderung der Koordination im Mehrebenensystem. *TATuP - Zeitschrift des ITAS zur Technikfolgenabschätzung*, *22(2)*, 48–55.

Ostrom, E. (1992). *Crafting institutions for self-governing irrigation systems*. San Francisco, Calif, Lanham, Md: ICS Press; Distributed to the trade by National Book Network.

Paradigm Environmental Strategies Pvt. Ltd. (2011). *Detailed Project Report for Nashik Municipal Corporation Waste to Energy Project.*

Pattberg, P., Widerberg, O., Isailovic, M., & Dias Guerra, F. (2014). Mapping and Measuring Fragmentation in Global Governance Architectures: A Framework for Analysis. Available at SSRN 2484513.

Peters, B. G. (2012). *Institutional Theory in Political Science* (3rd ed.). New York, N.Y: Continuum.

Platforma. (2011). *Decentralised Development cooperation – European perspectives*. Retrieved from http://www.platforma-dev.eu/files/upload/40/decentralised-development-cooperation--european-perspectives.pdf (19-02-2016)

Prins, G., Galiana, I., Green, C., Grundmann, R., Hulme, M., Korhola, A., Laird, F., Nordhaus, T., Pielke, R., Rayner, S., Sarewitz, D., Shellenberger, M., Stehr, N., Tezuka, H. (2010). The Hartwell paper: A new direction for climate policy after the crash of 2009. [Oxford, Eng.]: Institute for Science, Innovation & Society, University of Oxford.

Putnam, R. D. (2000). *Bowling alone: The collapse and revival of American community*. New York: Simon & Schuster.

Putnam, R. D. (Ed.). (2002). *Democracies in flux: The evolution of social capital in contemporary society*. Oxford, New York: Oxford University Press.

Putnam, R. D., Feldstein, L. M., & Cohen, D. (2003). *Better together: Restoring the American community*. New York: Simon & Schuster.

Putnam, R. D., Leonardi, R., & Nanetti, R. (1994). *Making democracy work: Civic traditions in modern Italy*. Princeton, N.J: Princeton University Press.
Qi, Y., Ma, L., Zhang, H., & Li, H. (2008). Translating a Global Issue Into Local Priority: China's Local Government Response to Climate Change. *The Journal of Environment & Development*, *17*(4), 379–400. doi:10.1177/1070496508326123
Ralston, H. A. (2013). *Subnational Partnerships for Sustainable Development. Transatlantic Cooperation between the United States and Germany. New Horizons in Environmental Politics series*: Edward Elgar Publishing Limited.
Revi, A. (2008). Climate change risk: an adaptation and mitigation agenda for Indian cities. *Environment and Urbanization*, *20*(1), 207–229. doi:10.1177/0956247808089157
Riddell, R. (2007). *Does foreign aid really work?* Oxford, New York: Oxford University Press.
Risse-Kappen, T. (1995). Bringing Transnational Relations Back In: Introduction. In T. Risse-Kappen (Ed.), *Bringing Transnational Relations Back In: Nonstate Actors, Domestic Structures and International Institutions* (p 3–33). Cambridge.
Robinson, J. (2002). Global and world cities: a view from off the map. International Journal of Urban and Regional Research, 26(3), 531–554. doi:10.1111/1468-2427.00397
Robinson, J. (2006). *Ordinary cities: Between modernity and development*. London [u.a.]: Routledge.
Robinson, J. (2011). Cities in a World of Cities: The Comparative Gesture. *International Journal of Urban and Regional Research*, *35*(1), 1–23. doi:10.1111/j.1468-2427.2010.00982.x
Rose, R. (1993). *Lesson-drawing in public policy: A guide to learning across time and space*. Chatham, N.J: Chatham House Publishers.
Rosenzweig, C., Solecki, W. D., Hammer, S. A., & Mehrotra, S. (2010). Cities lead the way in climate-change action. *Nature*, *467*, 909–911.
Said, E. W. (1994). Culture and imperialism. London: Vintage.
Sakhalkar, G. M. (2007). *Tram Allignment Final. PCMC, PMC (20-03-07)*. Pune.
Salisbury, R. H. (1969). An Exchange Theory of Interest Groups. *Midwest Journal of Political Science*, *13*(1), 1–32. doi:10.2307/2110212
Satterthwaite, D. (2008). Cities' contribution to global warming: notes on the allocation of greenhouse gas emissions. *Environment and Urbanization*, *20*(2), 539–549. doi:10.1177/0956247808096127
Satterthwaite, D. (2009). The implications of population growth and urbanization for climate change. *Environment and Urbanization*, *21*(2), 545–567. doi:10.1177/0956247809344361
Schreurs, M. A. (2008). From the Bottom Up: Local and Subnational Climate Change Politics. *The Journal of Environment & Development*, *17*(4), 343–355. doi:10.1177/1070496508326432
Schreurs, M. A. (2010). Multi-level Governance and Global Climate Change in East Asia. *Asian Economic Policy Review*, *5*(1), 88–105. doi:10.1111/j.1748-3131.2010.01150.x
Schröder, H., & Bulkeley, H. (2009). Global cities and the governance of climate change: what is the role of law in cities? *Fordham Urban Law Journal*, *36*(2), 313–359.
Schwedler, H.-U. (2011). *Integrated Urban Governance. The way forward. Commission 3. Manual*. Retrieved from http://www.stadtentwicklung.berlin.de/internationales_eu/staedte_regionen/download/projekte/metropolis/C3_Manual_barrier_free.pdf (19-02-2016)
Sharma, D., & Tomar, S. (2010). Mainstreaming climate change adaptation in Indian cities. *Environment and Urbanization*, *22*(2), 451–465.
Sippel, M. (2007). *CDM im Rahmen von Nord-Süd-Städtepartnerschaften – Potenzial zur Reduktion von Transaktionskosten?: Promotionsschrift zur Erlangung des akademischen Grades eines Doktors der Wirtschaftswissenschaften der Universität Flensburg*.
Sozialdemokratische Partei Deutschlands, Sozialdemokratische Partei Deutschlands, Landesorganization Bremen Bündnis 90/Die Grünen Landesverband Bremen. (2011). *Vereinbarung zur Zusammenarbeit in einer Regierungskoalition für die 18. Wahlperiode der Bremischen Bürgerschaft 2011-2015*.

Statz, A. & Wohlfahrt, C. (2010). *Kommunale Partnerschaften und Netzwerke. Ein Beitrag zu einer transnationalen Politik der Nachhaltigkeit.*: Heinrich Böll Stiftung. *Schriften zur Demokratie. Band 20.*

Stern, N. (2007). *The Economics of Climate Change: The Stern Review.* Cambridge.

Stockmann, R., Menzel, U., & Nuscheler, F. (2010). *Entwicklungspolitik: Theorien - Probleme - Strategien. Lehr- und Handbücher der Politikwissenschaft.* München: Oldenbourg.

Stone, D. (2000). Non-Governmental Policy Transfer: The Strategies of Independent Policy Institutes. *Governance, 13*(1), 45–70. doi:10.1111/0952-1895.00123

taz. Die Tageszeitung (2009, August 24). „Man wird ungeduldiger". Retrieved from http://www.taz.de/!39519/

Tews, K. (2008). Vom Erfolg anderer lernen. Policy-Transfer und seine Voraussetzungen. In C. Fischer (Ed.), *Stromsparen im Haushalt. Trends, Einsparpotenziale und neue Instrumente für eine nachhaltige Energiewirtschaft* (p 79–89).

The Economist (2014, September 20). Curbing climate change: The deepest cuts. Retrieved from http://www.economist.com/news/briefing/21618680-our-guide-actions-have-done-most-slow-global-warming-deepest-cuts (19-02-2016)

Tjandradewi, B. I., & Marcotullio, J. (2009). City-to-city networks: Asian perspectives on key elements and areas for success. *Habitat International, 33*(2), 165–172.

Tocqueville, A. d. (1987). *Über die Demokratie in Amerika. Manesse Bibliothek der Weltgeschichte.* Zürich: Manesse.

Tucker, E. (2010). Towards a more rigorous scientific approach to social measurement. Considering a grounded indicator approach to developing measurement tools. In G. Walford, E. Tucker, & M. Viswanathan (Eds.), *Handbook of measurement* (p 313–335). [S.l.]: Sage Publications.

United Cities and Local Governments. (2007). Press kit.

United Nations, Department of Economic and Social Affairs, Population Devision. (2014). *World Urbanization Prospects. The 2014 Revision.* Retrieved from http://esa.un.org/unpd/wup/default.aspx (19-02-2016)

United Nations Development Programme. (2000). *The Challenges of Linking.* New York.

United Nations Development Programme. (2009). *Charting A New Low-Carbon Route To Development United Nations Development Programme. A Primer on Integrated Climate Change Planning for Regional Governments.*

United Nations Framework Convention on Climate Change. (2015). Adoption of the Paris Agreement. Retrieved from https://unfccc.int/resource/docs/2015/cop21/eng/l09r01.pdf (19-02-2016)

United Nations World Commission on Environment and Development. (1987). Our Common Future / Brundtland Report. Retrieved from http://www.un-documents.net/our-common-future.pdf (19-02-2016)

Van der Pluijm, R., & Melissen, J. (2007). *City diplomacy: The expanding role of cities in international politics. Clingendael diplomacy papers: Vol. 10.* The Hague: Netherlands Institute of International Relations "Clingendael".

Van Ewijk, E., & Baud, I. (2009). Partnerships between Dutch municipalities and municipalities in countries of migration to the Netherlands; knowledge exchange and mutuality. *Habitat International, 33*(2), 218–226.

Wagner, C. (2006). *Das politische System Indiens.* Wiesbaden: VS Verlag für Sozialwissenschaften.

Walker, J. L. (1974). Performance Gaps, Policy Research, and Political Entrepreneurs: Toward a Theory of Agenda Setting. *Policy Studies Journal, 3*(1), 112–116. doi:10.1111/j.1541-0072.1974.tb01136.x

Walker, J. L. (1981). The diffusion of knowledge, policy communities and agenda setting: The relationship of knowledge and power. In J. E. Tropman, M. J. Dluhy, & R. M. Lind (Eds.), *Pergamon policy studies on social policy. New strategic perspectives on social policy* (p 75–96). New York: Pergamon Press.

Ward, K. (2008). Editorial—Toward a Comparative (Re)turn in Urban Studies? Some Reflections. *Urban Geography*, *29*(5), 405–410. doi:10.2747/0272-3638.29.5.405

Wolman, H., & Page, E. (2002). Policy Transfer among Local Governments: An Information-Theory Approach. *Governance*, *15*(4), 477–501. doi:10.1111/1468-0491.00198

Yin, R. K. (2009). *Case study research: Design and methods* (4th ed.). Los Angeles, Calif: Sage.

Young, O. R. (2002). *The institutional dimensions of environmental change: Fit, interplay, and scale*. Cambridge, Mass: MIT Press.

Yuen, B. K. P., & Kong, L. (2009). Climate Change and Urban Planning in Southeast Asia. *S.A.P.I.EN.S*. (2.3). Retrieved from http://sapiens.revues.org/881 (19-02-2016)

9 Appendix

9.1 List of Interviews

	Cooperation	Interview Partner	Stakeholder Group	Date
1		AFG (Association of Friends of Germany), Pune	Partnership association	26-11-2012
2				25-09-2013
3		Arbutus, Pune	NGO	09-12-2012
4				25-09-2013
5	Pune-Bremen			08-10-2013
6		BORDA (Bremen Overseas Research and Development Association), Bremen	NGO	11-09-2014
7				15-09-2014
8		Forum Städtesolidarität Bremen – Pune, Bremen	Partnership association	27-02-2013
9				01-10-2014
10		Journalist, Pune	Media	18-09-2013
11		LAFEZ (Bremen Landesamt für Entwicklungszusammenarbeit), Bremen	City Administration	27-02-2013
12		MAM (Maharashtra Arogya Mandal), Pune	NGO	19-09-2013

	Cooperation	Interview Partner	Stakeholder Group	Date
13		PMC (Pune Municipal Corporation), Pune	City Administration	29-11-2012
14				24-09-2013
15		Terre des Hommes, Pune	NGO	24-09-2013
16		Urban transport planner, Pune/Pimpri Chinchwad	Business	11-10-2013
17		Hamburg Wasser	Public Water Utility	01-03-2013
18				16-11-2012
19				05-12-2012
20				07-12-2012
21		GIZ, Nashik/Delhi	International Cooperation	30-09-2013 (1)
22	Nashik-Hamburg			30-09-2013 (2)
23				03-10-2013
24				27-11-2013
25				08-09-2014
26		Nashik MC (Nashik Municipal Corporation), Nashik	City Administration	05-12-2012
27				30-09-2013
28	Nagpur-Freiburg	Freiburg city administration (City of Freiburg, Department of Environmental Protection), Freiburg	City Administration	07-08-2012

9.1 List of Interviews 257

	Cooperation	Interview Partner	Stakeholder Group	Date
29				13-10-2011
30				08-08-2012
31		ICLEI, Nagpur/Delhi/Freiburg	City Network	31-10-2012
32				20-11-2012
33				05-12-2013
34				26-09-2014
35		Nagpur MC (Nagpur Municipal Corporation), Nagpur	City Administration	19-11-2012
36				14-10-2013
	Additional interviews, conducted during exploratory phase	Interview Partner	Stakeholder Group	Date
37		Berlin Senate, Department for Urban Development and the Environment, Berlin	City Administration	17-09-2012
38				19-09-2012
39		Bündnis 90/Die Grünen Berlin, Speaker for Climate Protection and Energy Policy, Berlin	Political Decision Maker	13-09-2012
40		BUND (Bund für Umwelt und Naturschutz Deutschland Freiburg), Freiburg	NGO	07-08-2012
41		Chargée de projet Service Agenda 21 - Ville durable, Geneva	City Administration	06-07-2012
42		Climate Alliance, Frankfurt a.M.	City Network	19-01-2012

	Cooperation	Interview Partner	Stakeholder Group	Date
43		Coimbatore Municipal Corporation, Coimbatore	City Administration	13-12-2012
44		Ecologic Institute, Berlin	Research	12-09-2012
45		Esslingen Coimbatore Association, Esslingen	Partnership Association	17-12-2012
46		Europäische Akademie, Berlin	Research	25-09-2012
47		Former Berlin Senator for Senatorin für Health, Environment and Consumer Protection, Berlin	Political Decision Maker	09-10-2012
48		GIRT (German-Indian Round Table Freiburg), Freiburg	Business	26-10-2012
49		Goethe Zentrum, Coimbatore	NGO	17-12-2012
50		Hanover Directorate of Environmental Affairs, Hannover	City Administration	17-01-2012
51		NIUA (National Institute of Urban Affairs), Delhi	Research	09-11-2012
52		Northern Virginia Regional Commission, Northern Virginia	Environmental Planner, Research	27-07-2012
53		Precocious Energytech Pvt. Ltd., Nagpur	Business	19-11-2012
54		School of Planning and Architecture, Delhi	Research	01-11-2012
55		TARU, Gurgaon	City Consultant	15-11-2012
56		Technische Universität, Berlin	Research	04-09-2012

Table 36: List of Interviews

9.2 Summary

The slow progress of the international UNFCCC climate negotiations has led policymakers and the researcher community to focus on subnational and transnational climate protection initiatives. In particular the role of cities and their potential to experiment with alternative or complementary mechanisms to national and international efforts in addressing global warming has been the subject of increasing discussion.

This study offers a highly significant and topical contribution to the research field of urban climate governance by addressing persisting research gaps in urban climate collaboration and policy transfer. Whereas experience of and knowledge about local greenhouse gas emissions reduction strategies is growing, there is still surprisingly little empirical and theoretical research available about city-level climate cooperation. Important research gaps remain on cooperation between cities from the Global South and Global North, different institutional designs and the role of private actors in urban climate collaboration. More research is also required on the conditions and processes leading to success or failure in city cooperation.

This study addresses these knowledge and research gaps in urban South-North climate cooperation through an analysis of four Indian-German urban partnership projects. Via a comparative case study analysis, the study investigates how and under which conditions German and Indian cities cooperate and learn from each other in the development of climate mitigation activities. Based on the development and application of a "grounded" index system the study explores conditions for success and failure in Indian-German partnership projects, using expert interviews and document analysis as methods of data collection. The study draws on a set of theoretical concepts (policy transfer, transnational climate governance networks, New Institutionalism, policy entrepreneur and social capital) that jointly exert great explanatory power, as they provide distinct but complementary perspectives with regard to the analysis of the conditions for transnational urban climate cooperation.

The case study findings demonstrate that transnational urban South-North cooperation on low carbon transitions is possible and is emerging in various institutional forms. In addition to more traditional twinning arrangements, cities also connect in ad hoc and project-oriented partnerships, which can be driven by external moderators such as transnational municipal networks and international cooperation agencies.

The data points to key challenges that remain in transnational urban partnerships. A first major challenge faced by such collaboration is widening the scope beyond micro-level and demonstration projects. A second key challenge is ensuring the post-project sustainability of transnational urban partnerships. The comparative case study analysis reveals a trade-off between these two challenges. The transnational urban partnerships analysed tend to take the form of either a bottom-up approach, based on social capital, driven by partnership entrepreneurs and leading to long-term collaboration on a micro level; or they are induced in a top-down manner, driven by external partnership moderators with close links to state institutions, and focussing on the implementation of a single project and thereby lacking a long-term perspective. The study concludes that improving multi-level governance coordination is an important prerequisite to addressing the challenges posed by bottom-up versus top-down approaches to the development of transnational urban North-South partnerships. It recommends focussing on institutionalising collaboration by setting up permanent local partnership offices as a crucial step towards better connecting partnership entrepreneurs, who are important facilitators for horizontal governance coordination, and external moderators, who are instrumental in providing vertical governance coordination. The study further recommends establishing and strengthening links with existing networks that foster urban exchange and learning, such as transnational subnational state partnerships, transnational municipal networks, and frameworks provided by national and supranational governments, international NGOs, donors or research projects.

A third key challenge of transnational urban partnerships is achieving mutuality, and the perception of it, in relationships between partners from the Global South and North. The results of this study suggest that a major variable for the extent of mutuality reached in urban North-South partnerships is the degree to which both partner cities pursue their own vested interests within the initiatives and thereby have a stake in the projects' successful realisation. To achieve this, partnerships are advised to systematically analyse and incorporate economic, social, and environmental co-benefits into the design of climate mitigation partnership projects. The study also advises that lessons should be drawn from the Pune-Bremen partnership which has experimented with shared funding responsibilities in joint projects.

The study closes by outlining the academic implications of the study; discussing conclusions for the further development and application of the theoretical concepts and methodological tools applied; suggesting areas for future research; and reflecting upon the difficulties in addressing the Northern bias in urban studies.

9.3 Zusammenfassung

Aufgrund des langsamen Fortschritts der internationalen UNFCCC Klimaverhandlungen richtet sich der Blick von politischen Entscheidungsträgern und der Forschung zunehmend auf subnationale und transnationale Klimaschutzaktivitäten. Vor allem die Rolle von Städten und ihr Potenzial, mit alternativen und komplementären Ansätzen zu nationalen und internationalen Klimaschutzanstrengungen zu experimentieren, werden zunehmend diskutiert.

Diese Studie leistet einen wichtigen und aktuellen Beitrag zum Forschungsgebiet der urbanen Klimaschutz-Governance, indem sie Forschungslücken im Bereich der transnationalen urbanen Zusammenarbeit adressiert. Während Erfahrungen und empirisches Wissen über lokale Strategien zur Reduktion von Treibhausgasen steigen, existiert überraschend wenig empirische und theoretische Forschung zu Klimaschutzkooperationen zwischen Städten. Forschungsbedarf besteht insbesondere zur Zusammenarbeit zwischen Städten aus dem Globalen Süden und Norden sowie mit Blick auf Faktoren, die Erfolg und Misserfolg in solchen Partnerschaften bedingen.

Diese Studie adressiert die Wissens- und Forschungslücken im Bereich urbaner Süd-Nord Klimaschutzkooperation über eine Analyse von vier Fallstudien zu indisch-deutschen Partnerschaftsprojekten. Mittels einer komparativen Fallstudienanalyse untersucht die Studie, wie und unter welchen Bedingungen deutsche und indische Städte in der Entwicklung von Aktivitäten kohlenstoffarmer Entwicklung zusammenarbeiten und voneinander lernen. Basierend auf ExpertInneninterviews und Dokumentenanalyse als Datenerhebungsmethoden und der Entwicklung und Anwendung eines „grounded" Index Systems exploriert die Studie Erfolgsbedingungen, Herausforderungen und Hemmnisse in indisch-deutschen Partnerschaftsprojekten.

Die Studie bezieht dabei theoretische Konzepte mit komplementären Perspektiven und Erklärungsansätzen für transnationale urbane Klimakooperation ein (Politiktransfer, transnationale Klimaschutzgovernance, New Institutionalism, Policy Entrepreneur und Sozialkapital).

Die Fallstudienergebnisse demonstrieren, dass transnationale urbane Süd-Nord Zusammenarbeit im Bereich Klimaschutz möglich ist und bereits in verschiedenen institutionellen Ausprägungen existiert. Neben Kooperationen in traditionelleren Städtepartnerschaften arbeiten Städte zunehmend auch in ad hoc entwickelten, projektorientierten Prozessen zusammen, welche häufig über externe ModeratorInnen wie beispielsweise transnationale Städtenetzwerke oder staatliche Organisationen für Internationale Kooperation initiiert und angeleitet werden.

Die Ergebnisse verweisen darüber hinaus auf drei wesentliche Herausforderungen, mit denen transnationale urbane Partnerschaften konfrontiert sein können.

Die erste Herausforderung ist die Entfaltung von Wirkungskraft jenseits von kleinen Pilot- und Demonstrationsprojekten. Eine zweite Schlüsselherausforderung ist die Sicherstellung von langfristigen Projektbeziehungen in urbanen Partnerschaften. Die vergleichende Fallstudienanalyse zeigt dabei Zielkonflikte zwischen Wirkungskraft und Langfristigkeit in Partnerschaftsprojekten auf. Die untersuchten urbanen Partnerschaften wurden entweder bottom-up aufgebaut, basierend auf Sozialkapital, angetrieben durch einen lokalen Policy Entrepreneur, mit kleinen Projekten und Langfristperspektive. Oder sie wurden über einen top-down Ansatz von externen ModeratorInnen mit engen Verbindungen zu staatlichen Institutionen initiiert, mit Fokus auf individuelle Projekte ohne Ziel einer langfristigen Partnerschaft. Die Studie schlussfolgert, dass eine bessere Mehrebenen-Governance Koordination eine wichtige Voraussetzung darstellt, um die Stärken von bottom-up entwickelten Partnerschaften (Langfristigkeit) und top-down induzierten Kooperationen (Wirkungskraft) zusammenzuführen. Eine zentrale Handlungsempfehlung ist die Etablierung von permanenten Partnerschaftsplattformen zur Stärkung des Austauschs und der Zusammenarbeit zwischen lokalen Partnerschaftsentrepreneuren (als Schlüsselakteure für horizontale Governancekoordination) und externen Partnerschaftsmoderatoren (als wichtige Vermittler für vertikale Governancekoordination). Weiterhin empfiehlt die Studie, dass transnationale urbane Partnerschaften stärker die Kapazitäten von bestehenden inter- und transnationalen Netzwerken und Plattformen für urbanen Erfahrungsaustausch nutzen sollten, beispielsweise transnationale Partnerschaften zwischen Bundesländern, Städtenetzwerke sowie Förderprojekte von Regierungen, internationalen NGOs, Stiftungen oder Forschungsprojekten.

Als dritte Schlüsselherausforderung identifiziert die Studien die Überwindung von einseitigen Partnerschaftsbeziehungen in transnationalen urbanen Nord-Süd-Projekten. Die Fallstudienergebnisse lassen darauf schließen, dass mehr Wechselseitigkeit in urbanen Klimaschutz-Partnerschaften durch eine systematische Adressierung der ökonomischen, ökologischen oder sozialen Co-Benefits in beiden Partnerstädten erreicht werden kann. Auch die Teilung von finanziellen Verantwortlichkeiten in Partnerschaften kann ein Mittel zur Stärkung von Nord-Süd Paritäten darstellen, wie die Partnerschaft Pune-Bremen demonstriert.

Abschließend werden die akademischen Implikationen der Studie erörtert. Dabei werden Schlussfolgerungen für die weitere Entwicklung und Anwendung der theoretischen Konzeption und methodischen Instrumente diskutiert, Empfehlungen für zukünftige Forschung ausgesprochen und die Herausforderung der Adressierung von Nord-Süd-Disparitäten in der Studie reflektiert.

The manufacturer's authorised representative in the EU is Springer
Nature Customer Service Centre GmbH, Europaplatz 3, 69115 Heidelberg,
Germany. If you have any concerns regarding our products, please
contact ProductSafety@springernature.com

Printed and bound by CPI Group (UK) Ltd, Croydon, CR0 4YY
25/03/2026
02078190-0003